BY MEGAN KIMBLE

*City Limits: Infrastructure, Inequality, and
the Future of America's Highways*

*Unprocessed: My City-Dwelling Year
of Reclaiming Real Food*

CITY
LIMITS

CITY LIMITS

INFRASTRUCTURE, INEQUALITY, AND THE FUTURE OF AMERICA'S HIGHWAYS

MEGAN KIMBLE

CROWN
NEW YORK

Published in the United States by Crown, an imprint of the Crown Publishing Group,
a division of Penguin Random House LLC, New York.

CROWN and the Crown colophon are registered trademarks of
Penguin Random House LLC.

Library of Congress Cataloging-in-Publication Data

Names: Kimble, Megan, author.
Title: City limits / Megan Kimble.
Description: First Edition. | New York: Crown, [2024] |
Includes bibliographical references and index.
Identifiers: LCCN 2023034296 (print) | LCCN 2023034297 (ebook) |
ISBN 9780593443781 (hardcover) | ISBN 9780593443798 (ebook)
Subjects: LCSH: Express highways—United States. | Transportation—Planning—
United States. | Cities and towns—Growth—Environmental aspects. | Cities and
towns—Growth—Social aspects. | Roads—Location—Environmental aspects.
Classification: LCC HE355 .K56 2024 (print) | LCC HE355 (ebook) |
DDC 388.1—dc23/eng/20230817
LC record available at lccn.loc.gov/2023034296
LC ebook record available at lccn.loc.gov/2023034297

Printed in the United States of America on acid-free paper

Map copyright © Drue Wagner

crownpublishing.com

2 4 6 8 9 7 5 3 1

FIRST EDITION

Editor: Aubrey Martinson
Production editor: Evan Camfield
Text designer: Ralph Fowler
Production manager: Sarah Feightner
Managing editors: Allison Fox and Sally Franklin
Indexer: Ina Gravitz
Publicist: Mary Moates
Marketer: Chantelle Walker

For 3H

While the major portion of the city plan herein described deals principally with that area inside the City Limits, we have continually kept in mind the fact that the present City Limits line is only temporary. The interests of the city extend far beyond this imaginary line.

—*A City Plan for Austin, Texas,* 1928

CONTENTS

PART III

CITY
LIMITS

THE HIGHWAYS

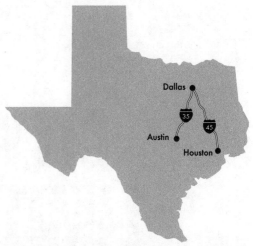

AUSTIN / I-35

The highway runs through the heart of Austin, dividing the east and west sides of the city. The I-35 Capital Express Central project would expand the highway from 12 to 20 lanes.

1. Escuelita del Alma
2. La Vista de Lopez
3. Community Land Trust home

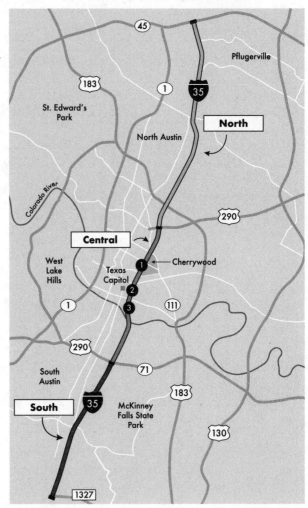

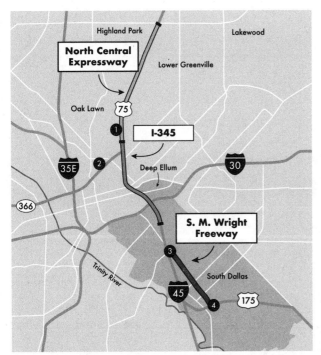

DALLAS / I-345

A 1.4-mile elevated highway forms the eastern edge of Dallas's downtown loop, connecting I-45 with the North Central Expressway. There is a campaign to remove the highway and replace it with a boulevard.

1. Freedman's Town/ North Dallas

2. Klyde Warren Park

3. Forest Theater

4. Dead Man's Curve

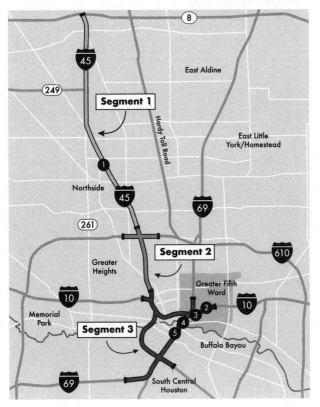

HOUSTON / I-45

The North Houston Highway Improvement Project, known as the I-45 expansion, will expand I-45 from downtown Houston north to Beltway 8 and rebuild and reroute the downtown loop, including I-10 and I-69.

1. Elda and Jesus Reyes

2. O'Nari Guidry

3. Modesti Cooper

4. Clayton Homes

5. The Lofts at the Ballpark

PART I

1

HOME

Houston
I-45

In the summer of 2019, Modesti Cooper went to the post office and checked her mail for the first time in months. She'd been living abroad for more than a decade, working as a civilian contractor for the U.S. military, moving from Kuwait to Afghanistan to Iraq—wherever the mission was. At work, she supported the military's technological capabilities: installing specialized software, revamping operating systems, recovering data, writing new computer programs. In her time off, she traveled—to Dubai, South Africa, Sweden—and collected art and souvenirs that she mailed back to Houston. When she had first moved abroad, after graduating with a master's degree in information technology from the University of Houston, she was in her early twenties and wanted to see the world. But as she approached thirty, she grew tired of living out of a suitcase and finding spiders in her twin bed. She wanted to put down roots.

So whenever she had a free moment, she looked for land. Sitting at her desk in Iraq, inside the reinforced walls of an Air Force base, she scrolled Google Earth, moving methodically up and down streets. She'd grown up in southwest Houston, near the Astrodome, but she wanted to live somewhere central. Eventually, she settled on the Fifth Ward, a quiet, historically Black neighborhood northeast of downtown. She bought a parcel of land, a small lot surrounded by trees on the corner of Nance and Grove Streets, and started designing her dream house from the ground up.

When she applied for construction permits from the City of Houston in 2016, someone told her that her house might be affected when the Texas Department of Transportation started construction on a long-planned rebuild of Interstate 10, which ran directly in front of her property. Construction would take years. It would be loud and some debris would probably settle in her yard. Oh, and there would need to be a twenty-five-foot buffer between the house and the highway frontage road. "I said, what am I going to do with all that space? Someone was just being funny and said, 'Put a pool in front of your house,'" she says. "That's how I ended up with a pool in front of my house." That's also how she ended up with a four-story home. Constrained to the back half of her property, she built up.

Now, whenever she had time off, instead of traveling to a new country, she went home to Houston. "A lot of people took pride in seeing me come here every day to pick up trash, keep the area clean, while it was being constructed," she says. Every few months, she'd visit and watch as her home took shape—the towering plywood skeleton gaining an angular roof, insulation and drywall, and then finally pale-blue stucco siding.

She came home in 2019 to finish moving her things into her new house, figuring that she'd work another year or two abroad to save money. But now, shuffled in among the bills and notices, she saw a letter from the Texas Department of Transportation, informing her that the state agency intended to acquire land to widen the highway that ran in front of her house as part of a project to expand Interstate 45 and rebuild and reroute Houston's downtown loop. The land TxDOT needed included the front half of her property.

Alarmed, she called the phone number on the letter. At first, it seemed as if TxDOT might work with her, would maybe skinny up the project slightly so that it didn't run through her brand-new house. She went into the TxDOT office to meet with staff in person.

"I was trying to explain to them, like, hey, you know, I was actually deployed to Iraq. So if there was any mail that came, I'd never seen it. I don't know what's going on," she says. The project had been in the works for fifteen years, they told her—how could she *not* know? Once she was there in person, sitting around a conference table with eight or nine TxDOT representatives, the tone shifted. "It was like sharks at the table," she says. They didn't need all of her land, she recalls them saying. "But you're not going to want to live in a house that when you come outside, there's a highway right there. So you might as well just give up the whole house." She could live in her house for a while, but eventually they would need to take it. "And I was like, well, how are you going to take my house? And they said, something called eminent domain."

TxDOT gave Modesti a printout titled "State Purchase of Right of Way." "Perhaps the first question that should be answered is 'Why does the government have the right to acquire private property?'" the brochure began. "Our successful existence in a democracy requires the development of public services to improve our way of life. . . . [A] government cannot provide services to its citizens without the right to acquire land." Indeed, eminent domain was enshrined in the U.S. Constitution's Fifth Amendment, which guaranteed that a person shall not be "deprived of life, liberty, or property, without due process of law; nor shall private property be taken for public use, without just compensation." For a government to legally seize private property, the use had only to be public and the owner justly compensated. Over twelve pages, the pamphlet explained exactly how the government would acquire Modesti's property, beginning with an appraisal and concluding, if a purchase price could not be agreed upon, with eminent domain proceedings, in which three disinterested landowners would be appointed to hold a hearing and determine the value of the property.

Modesti drove home, panicked. After that meeting, she decided not to return to her job in Iraq. Staying in Houston would give her the best chance of keeping her house. Modesti was raised by her grandmother; her parents had both died by the time she was seventeen. She didn't inherit wealth, didn't have anyone to bail her out when things got hard. One reason she'd stayed overseas for so long was that she had been able to bank almost everything she earned, a sacrifice for the future she was just now beginning to build for herself. Without anyone's help—without a safety net—she had found land and built herself a home. And now TxDOT wanted to take that away from her?

She stood on her fourth-floor balcony and looked west over the skyline of downtown Houston, underlined by the spaghetti junction of I-10 and I-69, cars streaming by on the tangled ribbons of gray. She loved the Fifth Ward. "It's a calm, cool, nice area," she says. "Besides the traffic, there's no violence, no noise. It's so quiet it's unbelievable. I had rockets and mortars and missiles blown over my head. To come home to peace of mind and say, okay, this is my forever home, you would imagine being satisfied just coming home to quiet." She was devastated by the prospect of losing the home she'd worked so hard to build. But she didn't think there was anything she could do about it.

2

DISTANCE

Austin
I-35

One morning in April 2020, the four members of the Texas Transportation Commission convene their monthly meeting by phone. The commission oversees the Texas Department of Transportation, which itself oversees more than eighty thousand miles of roadway, more than any other state in the country. The chairman of the commission, a banker from San Antonio named J. Bruce Bugg, Jr., reminds us in a pitch-perfect Texas drawl why we are here—separately, in our quarantined spaces. Today, the commission will consider allocating nearly $4 billion to expand Interstate 35 through central Austin.

"Keep in mind, no major improvements have been made to I-35 since 1974," Bugg says. That year, as the city's population approached 300,000, the highway was expanded into two levels, becoming the first double-deck interstate in Texas. No one on the call needs reminding that the population of the Austin metropolitan region recently surpassed 2 million people, or of the predictions that Austin's population will double yet again by 2040. Everyone has read the news reports, year after year, with the same headline: The Austin metropolitan region is the fastest growing in America. I-35 is "the main chokepoint in the major artery of Main Street Texas," Bugg says, with apparently no sense of irony that Main Streets have historically been characterized by pedestrian access to shops and services. It is the most congested stretch of highway in Texas, fourth in the nation for freight. "I will just close by asking, if not now, when?" he says.

It is an odd moment to conjure up such urgency to expand a highway. Earlier this morning, there was barely a ripple of brake lights on I-35. A month after the COVID-19 pandemic shuttered schools and businesses, forcing people to remain at home and connect through a computer instead of a car, it is hard to imagine our ominously gridlocked future, when the congested world we knew has disappeared so precipitously. But as the commissioners discuss the proposal in the awkward staccato of virtual communication, they remind their audience of the importance of the I-35 corridor to the state. Nearly half of Texans live and work along its spine. More than $750 billion worth of goods traverse it annually. "This is a huge economic engine that we should not be taking for granted," says Laura Ryan, the newest commissioner.

Before the commissioners vote on the funding, they open the meeting to public input. Early in the pandemic, we're still figuring out when to unmute what, and it takes a moment before the first caller, a man named Jay Blazek Crossley, speaks. He's with a nonprofit called Farm&City, and he is calling to tell the commissioners that the models they are using—the models that predict catastrophic traffic congestion as Austin booms—are wrong. These models assume that the future will look like the past, that everyone who moves to Austin will require a car to get where they need to go, and that where they are going will be very far from where they already are. Rebuilding I-35 is a once-in-a-century project. Maybe, Crossley says, we should entertain the idea that we might want the future to be different from the past.

Most of the roads that traverse Texas were built in the middle of the last century, when the Texas Highway Department, as the state agency was then known, was led by an engineer named Dewitt

Greer. When Greer retired in 1967, he warned of the catastrophic traffic congestion that would soon strangle cities. The state desperately and urgently needed to build more roads. "I do not believe mass transit is the answer in Texas," he said. "Texans are oriented to the use of the automobile. If we are to please the taxpayers . . . then we must develop more adequate thoroughfares in the urban areas."

State leaders have heeded that call. In 2014, Greg Abbott, now the Republican governor of Texas, ran for office promising to fix traffic in Texas cities. "A guy in a wheelchair can move faster than traffic on some roads in Texas," he said in a campaign ad, as he rolled slowly along the shoulder of an elevated Dallas highway next to stopped cars. Shortly after he was sworn in, Abbott appointed Bugg to lead the Texas Transportation Commission with the sole mandate to turn dirt to build roads. "I don't really know that much about transportation besides getting stuck behind the wheel of my car in traffic," Bugg told Abbott. Abbott responded, "That's all you need to know." And so Bugg had done as he was told, directing TxDOT to build more lanes to carry more cars. Between 2015 and 2023, Bugg dedicated more than $64 billion to expand highways through Texas cities.

I watched a live stream of this meeting from my home in Austin, less than a mile from Interstate 35. On cold, brittle nights, after the trees had lost their leaves, I could hear the highway humming in the distance. Before the pandemic grounded me at home, I drove on I-35 every day. It was a rickety, treacherous highway—the deadliest road in the city, with short on-ramps and confusing merges and too-fast access roads that careened through downtown.

I was born in Austin. My family lived in a three-bedroom house in West Austin on the edge of the Hill Country. Some of my first memories are of fire ants—the wrath of their small red bodies, the sharp surprise of being bitten. I remember the mugginess, how it felt sitting on the curb of our corner lot, wrapped in the thrumming green heat of Central Texas. When my parents bought their house on Rain Forest Drive in 1979, their neighborhood was brand-new, a remote development tucked into the stands of oak and cedar that rolled in from the west. Neither Loop 1 nor Highway 360 yet spanned the length of the city, as they soon would, crossing in a V-shaped intersection just south of the neighborhood. My dad often biked to the University of Texas, where he worked as a professor, pedaling down the quiet Rollingwood Drive, through Zilker Park, and along the trails that surrounded Town Lake. My parents, both vegetarians, shopped at Whole Foods, then just a tiny natural foods store on Tenth and Lamar that sold granola and coffee in bulk. In 1984, when my sister, Katie, was born, fewer than 400,000 people lived in Austin; by the time I arrived two years later, the population had already crept up to 450,000—a harbinger of the boom that was yet to come. It was one we would miss. Three years later, my dad accepted a job at Caltech and our family moved to Southern California, settling into the dusty foothills of the San Gabriel Mountains.

I learned how to drive on the highways of Los Angeles, steering— terrified—around the sharp corners of the Arroyo Seco Parkway, one of the country's oldest highways. One weekend a few months before I got my driver's license, I was driving with my dad on the 210 freeway toward Pasadena. "Let's say we're born with a certain amount of risk we get to spend in our lifetimes," my dad said. "Call it your risk bank." Every risk we took required a withdrawal. Some actions came with large onetime withdrawals—open-heart surgery, skydiving—while some were smaller debits, accumulating so slowly you couldn't see them, like breathing polluted air. "Do you know the thing that depletes your risk bank faster than any other activity?" my dad asked. "Driving. It is the most dangerous thing you'll ever do." This truth was oddly both alarming and empowering. It did not stop me from driving, but it put my safety in perspective. I have not been reckless or spent lavishly from my risk bank, but I have taken risks knowing that whatever I was doing was safer than driving—that whatever I did, the thing most likely to kill me was my car.

After college, I moved back to Los Angeles and commuted down-town, where I earned minimum wage working as an editorial assistant at the *Los Angeles Times.* To afford rent, I got a job as a private tutor, driving to homes spread across West Los Angeles to help high school students with their math and Spanish homework. Most days, I spent three or four hours driving. I didn't have a smartphone, so instead I studied traffic maps online before I left and hoped nothing would change while I was on my way. It always did. Snagged by traffic, I would arrive late and that lateness would compound, bumping into my next appointment. For two years, I lived with tensed shoulders, isolated and overwhelmed by what living in Los Angeles required of me.

Then, when I was twenty-four, I moved to Tucson, Arizona, and fell in love with the dry heat of the desert. I walked to work and biked most other places I needed to go. Most of my friends lived blocks from me, as we clustered near downtown, our proximity enabled by affordable rents. After I had spent years grinding on the freeways of Los Angeles, my relief was physical, all-encompassing. My life was contained within a three-mile radius, and so I rarely thought about transportation. It simply did not apply.

And then, after seven years in Tucson, I moved to Austin. When Katie had accepted a job as a professor at a liberal arts college just north of Austin, we laughed at her having ended up back there so long after our family had left. She and her husband bought a house and Katie got pregnant. Suddenly there was Ellie, manhandling my sister's iPhone to say hi to me. I didn't really want to leave Tucson, but I wanted to live near my family. So I moved. By the time I arrived in 2017, Austin was a fundamentally different city from the one my parents had left, with a population approaching a million and a booming tech economy. I got a job at a lifestyle magazine called *Austin Monthly* and moved into a creaky one-bedroom apartment in South Austin, dedicating nearly half my income to rent.

And my life once again filled up with driving.

My first assignment at the magazine was a ranking of the best breakfast tacos in Austin. I was given a list of two dozen to sample, with a deadline in three weeks. (It was, I joked to my sister on the twelfth taco, an Austin hazing.) As I drove to food trucks and brick-and-mortars in the same Honda Civic that had carried me across Los Angeles, I discovered a city built for cars, gridlocked by traffic. The breakfast tacos were delicious, but there was nowhere to wander them off, as the sidewalks kept ending, pedestrians an apparent after-thought in this liberal oasis. When I finished the breakfast taco story and tried biking to work, I found Austin's city streets to be treacher-ous, the bike lanes unprotected and always ending. The bus was no better—there wasn't even a route that passed by my office. A friend invited me over to his house for dinner, but he lived on the opposite side of town and the lonely drive through darkness felt insurmount-able. On weekends, I went out with co-workers, sat on patios under strings of globe lights, and listened to live music, but it felt like I was performing life in Austin instead of living it. Austin was supposed to be a progressive bastion in conservative Texas. But I didn't see that city. I just saw branding, a new idea projected onto old infrastruc-ture. I tried to get to know Austin, to experience the place that had captivated so many; mostly, though, it felt like I was stuck alone in my car for hours on end, driving to and from my sister's house, to and from my office, to and from the places and people I was writing about—all of which existed in separate spheres, disjointed from one another, uncollected into any kind of coherent community.

And I wondered, couldn't this be something better?

American cities have been carved up by roads, places paved over to create the arteries that allowed us to get someplace else. In 1924, as the automobile became ascendant, Herbert Hoover, then the secre-tary of commerce, declared that "locality has been annihilated, dis-tance . . . folded up into a pocket piece." Our cities reflect that annihilation of locality, the dissolution of the neighborhood unit as an anchoring force. As we built highways through—and away from—cities, we could get anywhere we wanted to go, but we found ourselves increasingly farther away from one another. Across the country, cities are sprawling farther and farther into previously undeveloped areas, enabled by the longer and wider highways that carry people home.

Conquering distance has come at enormous cost. There's the environmental cost: Transportation is the leading source of greenhouse gas emissions in the United States, accounting for a third of the country's total emissions. The majority of transportation emissions come from passenger cars and trucks. Even as our cars have become more fuel efficient, even as electric vehicles have trickled onto the market, emissions from passenger vehicles continue to climb. In Austin, the average driver emitted 12 percent more greenhouse gases in 2017 than they did in 1990, according to Boston University's Database of Road Transportation Emissions. In Dallas–Fort Worth, they emitted 27 percent more. Why? We are driving much more. "Suburban driving, including commuting, has been a major contributor to the expanding carbon footprint of urban areas," the project's lead researcher concluded. And Texas is the elephant in the room.

In 2018, the last time TxDOT did an analysis of the greenhouse gases generated on its roads, it found that on-road emissions in Texas account for nearly half a percent of total worldwide carbon dioxide emissions, more than some whole countries. Highway widenings in Texas alone have the potential to increase greenhouse gas emissions by more than thirty million metric tons by 2040.

There's the financial cost: It is very expensive to get around in a car. Today, the average family in Houston spends nearly 20 percent of their household income on transportation, compared with only 9 percent in New York City, where most people use transit. That cost disproportionately burdens low-income people. In 2015, a Harvard study found that in counties across the United States, the longer average commutes were, the harder it was for families to escape poverty. Indeed, commute times were the "single strongest factor" in predicting upward mobility.

And then there's the human cost, the roughly forty thousand people killed every year on our roads and the thousands more who get sick from breathing polluted air. Ten percent of the nation's traffic fatalities occur in Texas. In 2010, a panel convened by the Health Effects Institute looked at seven hundred worldwide studies of traffic-related air pollution. The panel found that within an exposure zone of roughly a quarter of a mile from a highway or major roadway, "the evidence is sufficient to support a causal relationship between exposure to traffic-related air pollution and exacerbation of asthma," particularly among children. A 2013 study found that air pollution related to road transportation contributed to fifty-three thousand early deaths every year.

In 1974, Yacov Zahavi, an Israeli engineer working for the World Bank, introduced the idea of a travel-time budget, which is the amount of time every day we're willing to devote to getting around. Whether in a rural village or a burgeoning metropolis, throughout human history the daily round-trip commute of any particular person has averaged out to just over an hour. Cities grew in concentric tree rings as we traveled farther but faster—on horseback, then in chariot and train, and, eventually, by car. In his book *Traffic: Why We Drive the Way We Do (and What It Says About Us)*, the journalist Tom Vanderbilt writes, "Higher speeds enable life to be lived at a scale in which time is more important than distance. . . . As societies, we have gradually accepted faster and faster speeds as a necessary part of a life of increasing distances."

This is the bargain of Austin and Houston and Dallas, the bargain of most American cities—increased distance predicated on speedy access. Across the country, as we built bigger and faster roads, we increased the sheer volume of land available for settlement. So we sprawled, pushing the limits of our cities farther and farther into the distance, assuming that speed would always be available to us.

But there is a growing movement to resist this form of development, to recalibrate our bargain with speed. Of all the costs of a world built on speed, the most dangerous one might be the cost to community. Highways divided our cities and dispersed us to far-flung suburbs, forcing us into cars to get anywhere worth going to. In the name of efficiency—of speed—we homogenized our cities, replicating placeless places across the varied wild landscapes of America. And as our communities flattened, we became increasingly anonymous and isolated within them, connected by wide rivers of concrete through which we hurtled at evolutionarily unthinkable speeds.

Once built, a highway can seem fixed, like a mountain—an unmoving feature of the urban landscape. But unlike a mountain, a highway breaks down. A highway is made of concrete, which degrades over time. Asphalt cracks. Steel rusts. Piers warp. Across the country, the urban highways we built so prolifically in the 1950s and 1960s are coming to the end of their physical life spans. Almost every city in the country has an aging highway coursing through its center that needs to be rebuilt. Mostly, state departments of transportation are dedicating billions of dollars to build those highways back wider and higher—despite decades of evidence that wider highways don't fix traffic.

In 1962, an economist named Anthony Downs published a paper titled "The Law of Peak-Hour Expressway Congestion." A native of Evanston, Illinois, Downs was a prolific researcher and spent his career as a consultant for think tanks, real estate developers, and the federal government. In the 1960s, he worked for the National Advisory Commission on Civil Disorders, which became known as the Kerner Commission after its final report, which Downs helped write, became a bestseller with its searing indictment of racism in America. The article he published in the journal *Traffic Quarterly* a few years before garnered significantly less attention.

"Recent experience on expressways in large U.S. cities suggests that traffic congestion is here forever," Downs began. It was only six years after President Dwight D. Eisenhower signed the Interstate Highway Act, catalyzing an unprecedented road-building boom. But no matter how many new roads were built, cars still slowed to a crawl at rush hour. This would seem to indicate, to frustrated commuters, "abysmally bad foresight by highway planners," he wrote. But to an economist, these results were so predictable as to be axiomatic. Downs's Law of Peak-Hour Traffic Congestion stated, "On urban commuter expressways, peak-hour traffic congestion rises to meet maximum capacity." Just as demand for goods fluctuated, demand for travel was not static. When presented with a wide-open expressway, commuters would flock to that expressway, eschewing other forms of travel. Offered this new ease of access, they might decide to move farther from their job or school, extending their commute, or take trips that might previously have been too costly, in time or money. Demand increases, outstripping the newly created supply. "We thus arrive at the paradoxical conclusion that the opening of an expressway could conceivably cause traffic congestion to become worse instead of better, and automobile commuting times to rise instead of fall!"

Researcher after researcher would replicate these findings: More lanes meant more traffic. Between 1993 and 2017, the hundred largest urbanized areas in the United States spent more than $500 billion adding new freeways or expanding existing ones. In those same cities, congestion increased by 144 percent, significantly outpacing population growth. And yet, somehow, the truth behind this basic phenomenon—so basic that in 2011 two economists would call it the Fundamental Law of Road Congestion—eluded transportation engineers across the country, who insisted, year after year, project after project, that this time they would be able to solve traffic congestion.

What is now Interstate 35 was once East Avenue, a four-lane boulevard split by a wide grassy median lined with manicured bushes and sprawling oak and pecan trees. Families picnicked in the shade between the north- and southbound lanes. On Saturdays, the median filled with vendors—farmers selling fresh produce out of the beds of pickup trucks, homemakers selling tightly wrapped tamales.

In 1946, Dewitt Greer, the state highway engineer, recommended to Austin's city council that a new superhighway pass through the city "in the vicinity of East Avenue." The City of Austin agreed to pay for the right-of-way needed to construct this new highway "after Greer frankly told them that 'if you don't want the East Avenue highway, say so, and we'll quit and spend our money elsewhere,'" reported *The Austin Statesman*. Two years before, a writer for *The Austin American*—the city's other daily newspaper—insisted that Austin needed a "super highway" if it was going to compete with other Texas cities. Dallas was already at work acquiring right-of-way for its Central Expressway, while Houston had begun building a superhighway from Galveston into the heart of the city, what would eventually become Interstate 45.

Two decades earlier, in 1928, Austin's first comprehensive plan allotted a six-square-mile "Negro district" on the east side of town, restricting municipal services and effectively prohibiting Black people from living west of East Avenue. The construction of the new Interregional Highway began in 1950, consuming East Avenue. The highway only reinforced this segregation, creating a physical barrier between Black and Hispanic communities and the heart of the city. In 1958, the Texas Highway Department started acquiring the right-of-way required to expand the Interregional Highway into Interstate 35. Over the coming years, the state would purchase 168 properties, consuming 156 acres of land. Mostly, that land was taken from the east side of the boulevard, where Black and Hispanic people lived and owned businesses.

There is essentially no urban highway in the United States that didn't unleash similar violence on Black and Hispanic people so that white people could get home faster. Over two decades, half a million homes were demolished and paved over. Today, urban highways continue to displace people. By the time the transportation commission was considering expanding I-35 in Austin, TxDOT was moving forward with a massive highway expansion in Houston that would consume nearly eleven hundred homes and more than three hundred businesses. "The effects of the project," TxDOT concluded, "would be predominantly borne by minority and low-income populations."

In Texas, as I talked to people displaced from their homes by these highways—then and now—I heard the same refrain repeated: You can't fight TxDOT. Whatever the state highway department wanted to do got done. You don't get to keep your house when the highway man wants it. You don't get to stay when they say go. The highway department spoke from a higher power, and that power was progress. It had engineers and experts, and traffic models so complex no ordinary person could understand them, so when it said that a new highway would save everyone time and money, would create prosperity and bring about a brighter, safer future, who would challenge this vision? The highway planners knew better, and anyway, they had the time and the money to wait you out and wear you down if you disagreed. Maybe they would nudge the highway this way or that way, particularly if you happened to be white, but one way or another, the highway would get built.

But despite the omnipotence of the highway builders, people did resist. In the 1960s, as the highways that had been conjured as lines on maps emerged into the real world, people revolted. They were appalled by the enormous structures that split apart their neighborhoods—all the homes that were taken, the ceaseless noise and polluted air that suddenly surrounded them. People poured onto city streets, crowded city council chambers, and wrote letters to their congressional representatives. In cities across the United States, freeway fighters erased highway lines from maps before they could be built.

In Washington, D.C., the Emergency Committee on the Transportation Crisis, or ECTC, mobilized low-income Black people to fight the district's plan to weave twenty-nine miles of interstate highways through the city. After planners canceled a highway that would have cut through an upper-class white neighborhood and instead proposed building the ten-lane North Central Freeway through a Black neighborhood, the ECTC rallied around the slogan "no white man's road through black man's home." After a decade-long fight, the plan for the North Central Freeway was abandoned. In San Francisco, after tens of thousands of people protested the city's highway plan, the board of supervisors voted to cancel many of the proposed routes, including the completion of the double-deck Embarcadero Freeway along the waterfront. In Baltimore, a multiracial coalition of residents called Movement Against Destruction successfully killed an east-west expressway that would have demolished twenty-eight thousand housing units. In New York City, a writer named Jane Jacobs stopped a planner named Robert Moses from building a four-lane road through Washington Square Park.

In Austin, city leaders approved a plan in 1962 for an extensive inner-city highway network, rendered as bright red lines on a black-and-white street grid. One of the proposed highways ran along the Colorado River, paralleling the north shore of the city's beloved Town Lake. Another proposed expressway extended north from the Texas State Capitol along what is now Guadalupe Street, cutting directly through central Austin. People were horrified. The Heritage Society of Austin wrote to the Bureau of Public Roads in June 1965. "Plans are being made which would drastically change the character of a large portion of the city," wrote the group's secretary. "While recognizing the inevitability of growth and change, it is the concern of The Heritage Society of Austin that this change should not destroy those unique qualities of Austin. . . . Let us benefit from the experience of San Francisco whose citizens halted the extension of one of the major overhead throughways which was cutting across the heart of the city after it had been under construction and the people saw what it would do to their city."

Rex Whitton, the federal highway administrator, replied. "Automobiles and the highways over which they must operate are an inte-

gral part of America, but such highways need not and should not be
disruptive elements in community planning," he wrote. "Rather
they should be, and we feel that they are at present, being planned
and located carefully, so that they enhance and stabilize the communities which they serve. . . . I would urge that all those persons such
as yourself who have the best interests of the area at heart, work
closely with your State highway officials to assure that a useful and
beautiful highway is built."

Austinites did not want a beautiful highway. They wanted the
homes and buildings that were already there. They wanted their waterfront. People crowded city hall to register their opposition. "The
City Council found out Thursday that charting a route for Central
Expressway will be a difficult job," reported *The Austin Statesman* in
1967. "At least doing it and keeping most people happy will be difficult." The Central Expressway and the Town Lake Expressway were
never built.

Back at the Texas Transportation Commission's April meeting, J. Bruce
Bugg thanks Jay Blazek Crossley for his comments. "May I have a motion to approve the updates to the 2020 Unified Transportation Program?" he asks. Ryan makes the motion, and Bugg seconds it and calls
for a vote. The funding measure passes, 3–1.

What Bugg and the other commissioners did not yet know on
that warm spring morning was that a second wave of freeway revolts had already begun.

3

FUTURE

When automobiles first trundled onto American streets at the turn of the twentieth century, the so-called horseless carriages were dismissed as a fad, a luxury item for the wealthy. "Although its price will probably fall in the future, it will never, of course, come into as common use as the bicycle," *The Literary Digest* reported in 1899, the same year *The New York Times* derided "these newfangled vehicles" as "unutterably ugly." Many states required that self-propelled vehicles travel no faster than four miles an hour and be preceded by a man on foot waving a red flag. But as automobiles rolled off assembly lines at relatively affordable prices, more and more people were able to experience the exhilarating freedom of moving about untethered to a rail or trolley line. In 1908, the year the Model T came out, there were 200,000 motor vehicles registered in the United States. By 1928, there were nearly 25 million.

In 1936, Shell Oil hired the prestigious New York advertising agency J. Walter Thompson to design a national campaign to promote driving. The oil company wanted millions more cars on the road every year, each one guzzling up Shell's new "motor-digestible" gasoline. To complete the commission, the ad agency turned to a young industrial designer named Norman Bel Geddes, a high school dropout from Michigan who moved to New York to study art and soon became an acclaimed industrial designer, tackling everything from household appliances to airplane cabins. After Shell approached him, Bel Geddes decided that the biggest obstacle to getting more cars on the road was the roads themselves, which had become choked with traffic, cars bottlenecked at congested intersections, pedestrians struck in horrifying numbers. Bel Geddes imagined a future of frictionless travel, tall buildings replacing low-slung housing, freeing up space for smooth, wide roads, with overpasses and curved ramps gently stitching traffic together. He built what he called the Shell City of Tomorrow, a six-foot-long model of a section of midtown Manhattan redesigned around the automobile, with wide highways and gleaming skyscrapers. Photographs of the model appeared in magazines and newspapers across the country.

A year later, when planners for the 1939 New York World's Fair began soliciting proposals for exhibits that would embody the fair's theme—"The World of Tomorrow"—Bel Geddes jumped at the chance to expand his City of Tomorrow model. Shell wasn't interested in sponsoring a full exhibit; its ad campaign was working just fine. Neither was Chrysler. General Motors turned Bel Geddes down three times. But Bel Geddes was persistent, and he finagled a meeting with Alfred Sloan, GM's president and CEO. During his pitch, Bel Geddes stood behind a trio of executives at GM's Manhattan headquarters as he narrated an immersive journey through the World of Tomorrow, captured in sixteen charcoal drawings. "The exhibit demonstrates that the world, far from being finished, has hardly begun, and that the future we create today will create richer lives and greater opportunities tomorrow," he began. Sloan was captivated. His vice president—who had already turned down the pitch twice—was less enthralled. "It won't sell a single automobile," he said.

"It sells the future," Bel Geddes replied. "With the promise that every citizen can own a piece of that future for the price of a General Motors' automobile." Sloan approved Bel Geddes's plan on the spot, eventually spending $6.7 million—more than $145 million today—on the thirty-five-thousand-square-foot model called Futurama.

The New York World's Fair opened on a hot, stormy day in April 1939; by the time it closed a year and a half later, forty-four million people had streamed through its gates. Many headed straight to the *Highways and Horizons* exhibit and queued up outside the bright white building, waiting for hours before they finally got a reprieve from the summer sun and descended into the cool dim of the Futurama model. The University of Texas architecture professor Lawrence Speck describes the scene that greeted them: "An enormous map of the U.S. seemed to hover magically over visitors. The map highlighted major cities as well as geographical features, but its primary focus was on roadways. First, strips of light depicted the current roadways of 1939 in bright red. Then, lights demonstrated the congestion that was expected to occur on this network with the phenomenal growth in automobile ownership projected for 1960. A third shift in lights envisioned a new system of superhighways that would link the country in a logical, efficient web of interconnectivity."

Past the lobby, visitors strapped themselves into six-foot-tall blue-mohair capsule chairs that glided through the exhibit on a conveyor belt as synchronized audio piped into individual speakers, narrating the scene that unfolded below. Bel Geddes's model covered almost four-fifths of an acre, containing half a million tiny buildings, a million trees, and more than fifty thousand automobiles—ten thousand of which actually moved. It captured the entirety of the country's landscape, from rural country towns to mountain hamlets, tidy suburbs surrounding skyscraper-lined cities. Stitching these distinct spheres together were wide roadways streaming with tiny cars. After nearly sixteen minutes traveling across a motorized America, viewers arrived to hover over a downtown intersection in the city of 1960, consisting of a tidy street grid filled with low-slung buildings mixed with glass-clad skyscrapers. Two seven-lane highways crossed in the middle, one folding seamlessly over the other. "As visitors exited their sound chairs, eased onto a moving sidewalk, and then alighted on solid ground, they found themselves at that same intersection at full scale," Speck writes. "They had actually entered the world of tomorrow." They walked on elevated sidewalks, gawked at the buses and trucks speeding below, peered up at the skyscrapers, and wandered into a showroom displaying new General Motors cars. Half an hour later, as people walked out of the exhibit, they pinned to their suits and dresses a blue-and-white button with the words I HAVE SEEN THE FUTURE.

When *Highways and Horizons* opened, after a decade of economic depression, the future offered by the automobile would have been mesmerizing. The exuberance Americans experienced was proportional to the sense of progress the automobile embodied. America still had so much wide-open country; the automobile would finally allow people to spread out and prosper, each family in its own home and with its own yard. Frank Lloyd Wright had articulated a similar vision for this car-focused future four years earlier when he described a new development called Broadacre City. The new American community, he wrote, would be based on a citizen's "freedom to decentralize." Land would be redistributed, an acre for every family. "The traffic problem has been given special attention, as the more mobilization is made a comfort and a facility the sooner will Broadacres arrive," he wrote. Every person would have their own car. "Multiple-lane highways make travel safe and enjoyable."

Before the advent of the automobile, only the wealthiest families had access to life outside the crowded, disease-ridden city. The car would be the great equalizer, allowing families of more modest means to move to the country, facilitated by their easy commutes back to the city. It would also save the city itself from ruin, replacing the smelly, space-intensive horse with a strikingly efficient machine and cleaning up dirty city centers by paving them over with smooth streets. Once articulated, Futurama became a self-fulling prophecy as corporate giants like General Motors and Shell Oil lobbied for the public roadways that would support private car travel. A few years after *Highways and Horizons* opened, construction industries, oil companies, and car manufacturers united under the American Road Builders' Association, becoming one of the largest lobbies in the country. And people flocked to buy the machines that promised freedom, progress, and autonomy. Ransom Olds, the creator of the Oldsmobile, declared that "the automobile has brought more progress than any other article ever manufactured."

To get to the 1939 New York World's Fair, most motorists would have crossed the Triborough Bridge, a massive structure built by Robert Moses, New York City's powerful parks commissioner, over the East River, connecting Queens and Long Island—including Flushing Meadows, the site of the World's Fair—to Manhattan and the Bronx and locales farther west.

The Triborough had opened three years earlier to great fanfare. More than 200,000 people gathered to watch as President Franklin Roosevelt took the first trip across the seventeen-and-a-half-mile bridge span, returning to extol the virtues of public works in an advanced society. When Moses took the stage, he lauded the "great artery . . . not merely a road for automobiles and trucks—it is a general city improvement, reclaiming dead areas and providing for residence along its borders." A journalist assigned by The New York Times to cover the bridge's opening reported that the Triborough would give motorists "a new freedom from congestion, delay and discomfort."

That same summer, Moses opened three expansive new freeways on Long Island—the Grand Central, Interborough, and Laurelton—connecting Brooklyn with parks and beaches miles to the east. By that point, Moses had spent a decade building more than a hundred miles of new roads across the New York metropolitan area. Traffic would be solved "for generations," wrote one newspaper columnist. But then something peculiar happened. Despite all these new roads and bridges, traffic seemed to get worse, not better. Of course there was traffic, Moses responded. He was just getting started, building roads as quickly as he could to accommodate demand. He proposed building forty-five additional miles of roadway, a giant loop around Brooklyn and Queens, to ease this congestion. But some city planners hesitated, noticing a pattern. "Every time a new parkway was built, it quickly became jammed with traffic, but the load on the old parkways was not significantly relieved," writes Robert Caro in *The Power Broker: Robert Moses and the Fall of New York*. "If this had been the pattern for the first hundred miles of parkways, they wondered, might it not be the pattern for the next forty-five also?"

The same mysterious pattern was plaguing the Triborough, too. The new bridge span was intended to relieve some of the pressure on the Queensboro Bridge, four miles to the south. But four months after the Triborough opened, "the relief of the traffic load on the Queensborough Bridge has not been as great as expected," reported the bridge's chief engineer. Every month, it seemed, engineers had to revise their projections—the eight-lane bridge would carry eight million cars annually, and then nine million, and then months later, they were predicting it would carry ten million cars a year. And yet traffic hadn't decreased on the other four bridges that spanned the East River. Where were all these cars coming from? "If traffic between the Island and the city was a stream, something had suddenly opened the sluice gates much wider than they had ever been before—and the more the traffic experts studied the problem, the more difficult it was for them to avoid the conclusion that the something was the only new element in the situation—nothing other than the Triborough Bridge itself," Caro writes. "Somehow, in ways they did not even pretend to understand, the construction of this bridge, the most gigantic and modern traffic-sorting and conveying machine in the world, had not only failed to cure the traffic problem it was supposed to solve—but had actually made it worse."

In 1939, as Bel Geddes was putting the final touches on *Highways and Horizons* in New York City, in Washington, D.C., Thomas H. MacDonald, the director of the Bureau of Public Roads, published a report titled *Toll Roads and Free Roads*, which conceived of a "special system of direct interregional highways, with all the necessary connections through and around cities." Before *Highways and Horizons* opened, Roosevelt had hosted a "no black tie—very informal" dinner at the White House for Bel Geddes. (At the president's insistence, the West Hall was set aside for a model of *Highways and Horizons*.) Months later, Bel Geddes wrote to Roosevelt describing the exhibit's popularity; Roosevelt responded by appointing Bel Geddes to help advise the creation of the national highway network proposed by MacDonald.

The federal government had been in the business of building highways since 1916, when the first Federal Aid Road Act passed. But states shouldered most of the cost of highway construction, and the program was rural only, focused on connecting ranchers and farmers with growing markets in cities. In 1944, Congress passed the Federal-Aid Highway Act, which authorized and designated a forty-thousand-mile "National System of Interstate Highways." For the first time, the act specifically set aside money to build highways in urban areas. States and the federal government would split the cost evenly, a fifty-fifty split. Although the act clearly established interregional highways as a national priority, Congress was unwilling to spend much to build these highways, allocating only $1.5 billion over three years. So over the next decade, most of these roads remained aspirational, rendered as lines on a map.

In 1952, a five-star general named Dwight D. Eisenhower was elected president. Thirty-three years before, as a young lieutenant colonel, Eisenhower had set out from the White House with the Transcontinental Motor Convoy, which traversed the country to survey the state of the nation's roads. They were not good—the 3,250-mile drive took two months—and Eisenhower returned harrowed by the journey. Later, when Eisenhower served as commander of the Allied forces in Europe during World War II, he was inspired by the German Autobahn, noting that "the old convoy had started me thinking about good, two-lane highways, but Germany had made me see the wisdom of broader ribbons across land."

On February 22, 1955, Eisenhower spoke to Congress. "Our unity as a nation is sustained by free communication of thought and by easy transportation of people and goods," he began. Without the unifying forces of communication and transportation, the nation would be "a mere alliance of many separate parts." One in every seven Americans gained his livelihood from the nation's highway system, Eisenhower said. But roads were about more than unity and jobs. It was the height of the Cold War. "In the case of an atomic attack on our key cities, the road net[work] must permit quick evacuation of target areas, mobilization of defense forces and maintenance of every essential economic function," he said. "But the present system in critical areas would be the breeder of a deadly congestion within hours of an attack." The critical road system conceived a decade earlier would not be completed to reasonable efficacy within the next half century, Eisenhower said. State highway departments alone could not meet the need; they simply didn't have enough money. A national highway network was of national interest, and so the federal government should pay for it.

In 1956, Eisenhower signed the National Interstate and Defense Highways Act—known as the Interstate Highway Act—which dedicated $25 billion to build forty-one thousand miles of interstate highways across the country over thirteen years. It was the largest public works project ever attempted in American history. To incentivize states to participate, the federal government agreed to pay 90 percent of the cost of these new interstate highways. The Bureau of Public Roads would manage the overall program, but states would select routes and handle construction contracts.

To oversee the implementation of the interstate highway system, Eisenhower appointed General John S. Bragdon as special assistant to the president. Eisenhower met Bragdon, a native of Pittsburgh, at the U.S. Military Academy at West Point when they were in their early twenties. The two men had become friends, staying in touch throughout their respective military careers. Bragdon served in France during World War I, was awarded a Purple Heart, and returned to earn a master's degree in engineering at the Carnegie Institute of Technology. He spent three decades working as an engineer for the military, eventually directing wartime construction for the U.S. Army. He was straitlaced, a rule follower; in his military portrait, he bears a distinct resemblance to Mr. Rogers.

When Eisenhower hired Bragdon, the president tasked him with coordinating "as rapidly as possible" the planning and construction of public works across government agencies, with a special focus on the Bureau of Public Roads. Eisenhower wanted to be kept informed about what was being built and how much it cost. The president was adamant from the beginning: The interstate system should be self-financing. It would be paid for by taxes on gasoline, motor vehicles, and manufactured goods like tires. Although the federal government had been taxing gasoline for decades, with the passage of the Interstate Highway Act those taxes would be collected into the newly created Highway Trust Fund and earmarked exclusively for interstate construction.

In June 1959, Bragdon wrote to Eisenhower requesting a meeting. "Inquiry reveals that many cities wish to take extreme advantage of the very liberal 90-10 provisions to solve their local problems of congestion which may include commuter traffic and other local needs," he wrote. These urban highways were consuming a disproportionate share of the interstate budget—44 percent of federal dollars, even as they accounted for only 10 percent of the total mileage. Urban highways cost on average $4 million a mile to build, compared with only $880,000 a mile in rural areas. As cities clamored for congestion relief, the estimated cost of the total system had ballooned to a projected $36 billion. That $11 billion deficit was now the responsibility of the federal government, and the federal government didn't have the money to pay for it. Two weeks later, Eisenhower tasked his deputy with a broad review of the entire highway program, including the "policies, methods, and standards" in place that might prevent it from achieving national objectives. Priority should be given to cost-saving measures. "As specific conclusions and recommendations are developed, I expect them to be implemented," Eisenhower wrote.

The Interstate Highway Act had not included specific criteria for designating urban routes, even as it greatly increased the amount of money available to states to build those routes. Before he offered any recommendation, Bragdon wanted to know to what extent Congress thought federal money should be spent to solve congestion within cities, rather than facilitating connections between cities. More to the point: Should interstates be routed through cities or around them?

Bragdon asked the Bureau of the Budget and the Department of Commerce to investigate. A few months later, he received a report titled "Legislative Intent with Respect to the Location of Interstate Routes in Urban Areas." The agencies recognized that by substantially increasing the federal contribution for new road construction, Congress had created a "new policy problem" that it had failed to solve: "the extent to which urban connections of the Interstate system are designed to serve local vehicular transportation requirements." This problem traced its origins to two words tucked into the 1956 law, which read, "Existing highways located on an interstate route shall be used to the extent that such use is practicable, suitable, and feasible, it being the intent that local needs . . . shall be given equal consideration with the needs of interstate commerce." No one could agree what "local needs" meant, even those who wrote the statute. Did it mean that state highway planners should accommodate the preferences of local officials when routing interstates through cities? Did the phrase simply permit states to bring existing roads up to interstate standards, thus avoiding duplication? Or did it mean that interstate highways should somehow serve both local and national traffic? "How can you tell which deserves the interstate treatment and which deserves the urban?" Senator Prescott Bush asked in a 1955 hearing. "There's quite a difference."

In 1956, Eisenhower had appointed Bertram Tallamy to lead the Bureau of Public Roads, which was then nested within the Department of Commerce. Tallamy was an engineer and a disciple of Robert Moses; he would later tell Robert Caro, Moses's biographer, that "the principles on which the System was built were principles that Robert Moses taught him." And Robert Moses believed, often to a ruthless extent, that highways belonged in cities and anyone standing—or living—in the way of highways was standing in the way of progress. In 1954, Moses argued in front of the President's Advisory Committee on a National Highway System that new urban expressways "must go right through cities and not around them."

But Congress had not explicitly endorsed this proposal. "The intent of the language 'equal consideration' of 'local needs' cannot be clearly drawn from the legislative history," the report concluded. However, a "logical interpretation . . . would permit a more narrow construction than has been applied by the Bureau of Public Roads" under Tallamy. In other words, local needs—namely, solving urban congestion within cities—were subservient to the national one of connecting the country. Federal money should be spent on federal priorities, not local ones.

On April 6, 1960, Bragdon presented the findings of his nine-month-long investigation to President Eisenhower. It was a typical day for the president: Eisenhower went to the doctor and arrived at the Oval Office shortly after 8:00 A.M. He enjoyed a quick visit with his son, Major John Eisenhower, and then began his day's meetings. At 10:35 A.M., Bragdon arrived at his office, accompanied by Frederick Mueller, the secretary of commerce, and Tallamy. Bragdon recounted the impetus for the meeting—that certain "excessive" practices had been observed in the implementation of the interstate program, which had concerned the president. The concept of the program had departed from its original objective, which was to connect the country to serve national defense and interstate commerce, Bragdon said. The highway system had been co-opted to solve local traffic problems within cities—to the detriment of the overall program.

"We do not believe that Interstate highways should be directed towards the center of congestion, as is now the case," Bragdon said. "Nor do we believe that inner loops, which are primarily for the purpose of solving intra-city problems, should be a part of the Interstate system . . . More emphatically, we do not believe that the Interstate System is the vehicle for solving rush-hour traffic problems, or for local bottlenecks . . . Practically all the experts on the transportation problem of cities agree that rapid transit and mass transit systems are the solution. Rapid transit and mass transit, by means of subways, commuter trains, special bus lines, are the answers for solving this problem." Bragdon was an engineer, dispassionate and pragmatic. The problem of rush-hour traffic was one of simple geometry: Cars took up significantly more space than people did. In crowded cities, where space was at a premium, it was more efficient to move people than it was to move cars. But instead of increasing public transit, cities were actively destroying it to make space for automobile infrastructure. "In Los Angeles, where there used to be some 2,300 commuter trains daily, there are none now," Bragdon told Eisenhower. "Commuter traffic by rail in Philadelphia has been so cut into by the automobile, that the rail commuter service can operate only at a loss."

Bragdon suggested that Eisenhower speak to Congress to clarify the limits of the system and the intent as to local needs. In the meantime, Bragdon said, the Bureau of Public Roads should revise its criteria regarding the routing of interstate highways and make explicit that highways should bypass cities rather than penetrate them. The implementation of modified criteria would reduce the cost of the program by at least $5 billion. But more than that, it would force cities to undertake proper urban planning. "The highway plan should not be the central pattern around which a community develops," Bragdon told the assembled men. "The basic plan for all community development should be an economic growth and land use plan." Highways were but one form of transportation, he said, and transportation was but one way to plan a community.

Eisenhower's response to this presentation was captured in a memorandum of the meeting published a few days later. Eisenhower responded, apparently in frustration, that "the matter of running Interstate routes through the congested parts of the cities was entirely against his original concept and wishes; that he never anticipated that the program would turn out this way. . . . He was certainly not aware of any concept of using the program to build up an extensive intra-city route network as part of the program he sponsored. He added that those who had not advised him that such was being done, and those who had steered the program in such a direction, had not followed his wishes."

But Eisenhower's hands were tied, he said. Routes had already been assigned, funding committed to states. Indeed, Mueller and Tallamy insisted that if such criteria were implemented, "the States would rise up in arms." Promises had been made. And it was an election year. Eisenhower's vice president, Richard Nixon, was running against a handsome young senator from Massachusetts named John F. Kennedy. Eisenhower's secretary, Ann Whitman, recapped the day's meetings in her diary: "Secretary Mueller, Mr. Tallamy, General Bragdon and others in for a hearing on the roads program. I do not know the details but I gather General Bragdon wants something to be done now—and General Persons and others think it would be murder in an election year to move."

At 11:15, the meeting ended. Eisenhower had other commitments, including a meeting with the president of Colombia, who was in D.C. for an official state visit. Eisenhower would begin that meeting by complaining about his previous one, in which he was made aware of "certain misunderstandings on the part of the Congress regarding that portion of the cost to be borne by the municipalities through which the highways will pass."

Nothing more would come of Bragdon's report. No directive would be given to the Bureau of Public Roads to change course—to instruct states to focus federal funding on connecting the country, rather than the newly invented problem of urban congestion. Less than a month later, Eisenhower reassigned Bragdon to the post of Civil Aeronautics Board, and his job as coordinator of public works planning was liquidated. It's unclear if the president was trying to take care of his old classmate as he prepared to leave office or if the transfer was an attempt to distance his administration from Bragdon's report. "General Bragdon's last major assignment from the President, an over-all study and report on the federal-aid highway program . . . caused some consternation among federal officials and congressmen by recommending cutdowns on approved mileage, standards and plans, particularly for urban areas," reported the *Engineering News-Record*, a weekly trade magazine, in May 1960. "At a special White House meeting recently, the President and some of his responsible officials discussed with General Bragdon the tenor of the findings and recommendations to appear in the final report. The recommendations found little favor with them. As a result the report is being heavily discounted in White House circles. It is doubtful now that the President will ever release it."

4

BLIGHT

Houston
I-45

When O'Nari Burleson was born in 1948, Houston's Fifth Ward was a thriving working-class Black neighborhood just east of downtown. After the Civil War, a freedman named Toby Gregg established a church on Vine Street, on the banks of the Buffalo Bayou, and freed slaves settled in the sparsely populated area. As the railroad industry boomed in Houston, foundries and repair shops concentrated in the neighborhood, taking advantage of its proximity to the Houston Ship Channel. Creoles of color, fleeing floods in Louisiana, found jobs at the massive Englewood Yard of the Southern Pacific Railroad and built homes nearby, giving the neighborhood the nickname Frenchtown.

James and Alice Burleson had both grown up in rural East Texas, but they met and married in the Fifth Ward sometime in the 1930s and bought two adjacent lots on Bringhurst between Buck and Providence Streets. They built a pier-and-beam house with a living room, a kitchen, two bedrooms, and a bathroom. Alice was pregnant when James was deployed to World War II. He met his son, O'Nari's big brother, when he returned to Houston, five years later. By the time O'Nari was born, more than forty-four thousand people lived in the Fifth Ward. It was a tight-knit community; everyone knew everyone.

No one locked their doors; O'Nari's house didn't even have locks on the front door. Everybody had more or less the same stuff, nothing worth stealing. No one knocked, either. "They'd say, 'Hey, ho, how you doin'?' and walk on in," she says. Most everyone had a garden running alongside their house, with collards and cauliflower sprouting through the winter. O'Nari's house had a big front porch with a swing on it. "That used to be the neighborhood entertainment," she says. A bayou ran along the back of their property. "I wasn't scared of it, but it wasn't a place to play. It was swift water in it all the time."

Alice worked as a seamstress, sewing clothes for people in the neighborhood out of their home. After World War II, James became one of the first Black people in Houston to work for the U.S. Postal Service. The family didn't own a car, so James walked downtown every morning to the central post office. Everybody walked everywhere in those days. O'Nari and her mom often walked up to Lyons Avenue, the commercial and cultural heart of the community. You could get anything you needed on Lyons Avenue. There were barbershops and beauty salons, drugstores and medical clinics, shoe hospitals and tailors, restaurants and clubs. It was a self-contained community; before integration, there was no reason to leave.

O'Nari walked to school, too. She was ten or eleven years old, in fifth or sixth grade at Bruce Elementary, when she first heard rumors that a highway was going to be built through the Fifth Ward. It was hard to imagine as she walked past the tidy shotgun homes and bungalows, people rocking on porches in the shade of soaring pine trees. But by the time she started at E. O. Smith Junior High School, the highway was a fact. One day, her parents got a letter from the Texas Highway Department. The Burlesons' home was on land the agency needed to extend Interstate 10 west into downtown Houston. Families were given six months to find new housing. It was 1959, almost a decade before the Fair Housing Act would open up neighborhoods to nonwhite families. The Fifth Ward was one of only a few places in Houston where Black people could own homes and operate businesses. "I can remember my mama crying, my grandmother upset, my daddy upset," she says. O'Nari was barely a teenager, and suddenly her life was upended. "Everyone around me was devastated of having to uproot and move."

The highway happened quickly. Kids went to school and returned home to holes in the sky, the pine trees that shaded the community felled by construction crews. Homes disappeared overnight. Some were bulldozed, but many were moved. Giant trucks rumbled through the neighborhood in the darkness, idling while workers disconnected plumbing and gas lines, and then rolled away with a whole house in tow, leaving behind vacant pits.

A few blocks north of where O'Nari lived, past Lyons Avenue, a woman named Ella Morris woke up one night to a big racket. She got out of bed and looked out the front window of the two-bedroom house she lived in with her husband and two young children. There, in the middle of the dark street, was a house perched on the back of a flatbed truck. The house looked exactly like hers—a small wood-framed bungalow with a gable-roofed porch—but the exterior was painted canary yellow instead of sage green. Ella didn't venture off her street much; her husband built furniture while she raised their children. There was a grocery store up the block, and she sometimes walked around the corner to drink coffee on a friend's porch in the early morning. She didn't know anything about a highway coming to her neighborhood. All she knew was that the Fifth Ward was changing; strange things were happening like houses floating through the streets in the middle of the night. "I wonder where they're taking that house," she thought. She went back to bed. When she woke up the next morning, the house was in her backyard, eight feet from the rear door, where it would remain for the following six decades.

Many homes and families ended up in Kashmere Gardens, a neighborhood directly to the north of the Fifth Ward. After they got the letter from the Texas Highway Department, James and Alice bought a vacant lot in Kashmere Gardens and set to building yet another home from the ground up. They moved when O'Nari was thirteen years old, the summer before she started the ninth grade.

After they moved, James bought the family's first car; it was too far for him to walk to the post office in downtown Houston from their new neighborhood. Public transportation was erratic at best. "The buses would come when they wanted to," O'Nari says. White bus drivers sometimes stopped for Black riders, and sometimes they didn't.

Although she was zoned for Kashmere High School, O'Nari was determined to graduate from Phillis Wheatley High School, one of the largest and best Black high schools in the country. A Wheatley graduate named Barbara Jordan had recently become the third Black person licensed to practice law in the state of Texas and set up a practice on Lyons Avenue (within the decade, Jordan would become the first Black woman ever elected to Congress from the South). Dozens of O'Nari's new neighbors were also displaced Fifth Ward families. Everyone went to Wheatley—that's just what you did if you were from the Fifth Ward, she says. So early every morning, the kids would gather and walk nearly four miles back to the high school, crossing Cavalcade and Collingsworth and Lyons Avenue, cutting through backyards and threading along the alleyways between houses. To O'Nari, it didn't seem like that long of a walk. She had plenty of energy and lots to talk about. They would cross over the highway's route, which by then had been scraped clean and depressed, a muddy trench that the kids walked over on narrow, dangerous pedestrian bridges.

Sometimes, a boy named Curley Guidry would walk her home after school and then walk all the way back to his family's house on Orange Street, seven miles round-trip. He was a year older than her, tall and lanky, with a wry, wide smile. He was smitten and O'Nari knew it, but she didn't want a boyfriend just yet; she was still so young. Nevertheless, they spent hours together, walking across the neighborhood, lingering after football and baseball games. Curley taught her how to drive at Finnigan Park, across the street from Wheatley.

Before her family's home was bulldozed, O'Nari hadn't thought much about her future beyond getting married and having kids. But being forced to move had shifted something within her. When she lived in the Fifth Ward, she'd been a child—safe, cocooned in the familiar. Being uprooted from the Fifth Ward just as she leaned into adulthood forced her to consider her future, how she might support herself if she could no longer live with her parents. In 1964, when she was a sophomore in high school, Lyndon B. Johnson signed the Civil Rights Act, which opened new academic programs to Black students. One of those programs was vocational training, which Wheatley offered. So when an academic counselor suggested that O'Nari look into nursing, she enrolled, starting classes while she was still finishing her senior year.

Interstate 10 opened in the Fifth Ward on April 18, 1966. Officials marked the occasion with a ribbon cutting on the Gregg Street overpass, which the *Houston Chronicle* covered with a brief news story, reporting, "One of the most expensive 3.5 miles in the Interstate Highway System was opened in Houston today by Herbert C. Petry, chairman of the Texas Highway Commission." The stretch of highway, which cost $15 million, "is of the most modern design," Petry said at the event.

Jayne McCullough was a junior at Wheatley when the highway opened. She was a majorette with the Wheatley marching band, which had been asked to perform at the highway ribbon cutting. The band marched along the highway, the concrete clean and vacant of cars. Jayne wore a sparkly skirt and tights, dancing and twirling her baton at the front of the procession. The eight-lane highway stretched across what had been three city blocks in the heart of her neighborhood. She remembered, vividly, the single-story homes and soaring trees and cracked sidewalks that had been taken. Even as she smiled and twirled, Jayne was distraught, thinking, "This doesn't make sense. Why are we glorifying something that tore up our community? We're supposed to be celebrating and happy about this? No, this wasn't no happy celebration to me." Her friends were upset, too. But they were just teenagers: "We were young enough to feel some way about it, but you couldn't say nothing because we were just kids."

Curley biked to the ribbon cutting from his family's home, which overlooked the new highway. The highway had consumed most of Orange Street, but it had spared their house when it swung south. Their living room window had once looked out over a row of low-slung homes and porches, nearly identical to theirs. Curley and his younger brother, Henry, watched as those houses disappeared, one after another. Henry's paper route had disappeared, too, as the businesses on Stonewall Street shuttered, displaced by the highway. Now the view out their front window was concrete gray, a trench of vacant space. Only seven homes remained on their block. Curley was nineteen years old when the highway opened. It made him uneasy, the way it carved up his neighborhood and displaced his friends. But the highway had been sold as progress, and he wanted to go see what that progress looked like. So he got on his bike and pedaled a few blocks west, navigating around the newly dead-ended streets.

There was a big crowd on Gregg Street, big enough that he couldn't get very close to the action, so he hovered near the back. "It was all dignitaries and highway people," he says. Most of the people who lived in the neighborhood were at work. He watched as the Wheatley students marched and danced as a white man with big scissors cut a fluttering paper ribbon stretched across the road. Then he biked back home.

All told, 1,220 structures were wiped from the Fifth Ward, including eleven churches, five schools, and two hospitals.

The idea to use interstate highways to remove "slums" and "blight" originated in the 1939 report *Toll Roads and Free Roads,* published by the highway engineer Thomas H. MacDonald. In it, MacDonald suggested that urban highways could serve a dual purpose: move cars and clear slums. "Citizens with adequate income"—white citizens—were leaving their homes in the center of the city in favor of the suburbs, he wrote. "The motor vehicle itself is the primary cause of this phenomenon." The homes these citizens left behind "have descended by stages to lower and lower income groups. . . . Almost untenable, occupied by the humblest citizens, they fringe the business district, and form the city's slums—a blight near its very core!" The federal government had begun to acquire these homes "in batches in connection with its slum-clearance projects. . . . These acquisitions comprise one of the reasons for avoidance of delay in dealing with the problem of transcity highway connections and express highways."

The suburbs and highways were mutually reinforcing: The suburbs justified the highways, and the highways enabled the growth of the suburbs. After World War II, millions of veterans arrived home to a housing crisis—or rather, by arriving home, they created one. Because construction of all kinds had essentially halted during the Great Depression and ensuing world war, there simply weren't enough dwellings to go around. In 1946, Congress declared a national housing emergency to incentivize the construction of millions of homes needed to accommodate the growing population returning from war. Concurrently, the government created a ready market for new construction through the GI Bill, which made low-interest, zero-down-payment loans available to servicemen, who were quickly marrying and multiplying and looking for somewhere to settle.

In 1947, a U.S. Navy veteran named William Levitt saw the massive housing shortages across the country and wondered why homes weren't manufactured like cars. He bought seven square miles of potato fields on Long Island and built a neighborhood called Levittown using the same assembly-line techniques that Henry Ford had used to make the Model T affordable to Americans. He built 17,500 homes in just four years, selling two-bedroom homes to returning veterans for just $8,000, no down payment required. This pattern of development proliferated across the country. Developers bought cheap land on the fringes of cities. They felled trees and poured concrete, etched roads and dug sewer lines, framed houses and brought out buyers. Across the country, white residents qualified for home loans and moved out of congested city centers. Black and Hispanic families, meanwhile, were blocked from the financing required to buy these homes—they were denied federally backed mortgages—and excluded from the neighborhoods those homes were built in through racially restrictive covenants. And so they remained in increasingly hollowed-out city centers while white families hurried home to the suburbs on the highways that unfurled before them.

In 1944, a government report recommended that state and local authorities purchase needed lands for highways that would "aid in the efficient assembly and appropriate redevelopment of large tracts of blighted urban lands." As the Interstate Highway Act was being written, the American Association of State Highway Officials sent lobbyists to D.C. to influence the legislation. Alfred Johnson, the group's executive director, later recalled that "some city officials expressed the view in the mid-1950s that the urban Interstates would give them a good opportunity to get rid of the local 'n——rtown.'"

The "blight" that urban highway planners sought to eliminate had in fact been created by the federal government decades earlier. In the early 1930s—around the same time O'Nari's parents moved to the Fifth Ward—the Mortgage Rehabilitation Division of the Home Owners' Loan Corporation, or HOLC, sent an appraiser to Houston. The HOLC was created in 1933 to stabilize the housing market, which had been hit by a wave of foreclosures during the Great Depression. In addition to buying up existing mortgages, the HOLC started issuing new federally insured mortgage loans. To assess the riskiness of lending in certain neighborhoods, the HOLC sent appraisers to more than two hundred cities across the United States, where they toured neighborhoods, noting the age of houses and their condition. But appraisers also considered the racial and ethnic makeup of the people who lived in those houses and downgraded neighborhoods with higher concentrations of Black and Hispanic people. Standardized maps rated neighborhoods by color: green for the "Best," blue for "Still Desirable," yellow for "Definitely Declining," and red for "Hazardous." Red neighborhoods were "characterized by detrimental influences . . . undesirable population or an infiltration of it," according to HOLC criteria. Appraisers recommended that lenders "refuse to make loans in these areas or only on a conservative basis."

The Fifth Ward was shaded entirely red. Stair-stepping from McKee Street, just east of downtown, all the way to Sakowitz Street, roughly five hundred blocks were labeled hazardous. If most people who lived in the Fifth Ward thought of it as a thriving, self-contained community, the government did not agree.

In the late 1950s, when it came time to build interstates through American cities, city planners looked to these redlined neighborhoods as they drew their routes. In 2018, two researchers compared digitized redlining maps with interstate routes and found that redlined neighborhoods were three times more likely to have an interstate highway routed through them than the best-rated neighborhoods. "Interstate highways caused sharp and persistent declines in population and housing stock for decades after they were built, and these declines were in addition to those resulting from neighborhood poverty, racial segregation, white flight, or urban renewal policies," the researchers concluded.

According to Eric Avila, an urban planning professor at the University of California, Los Angeles, between 1956 and 1966 highway construction demolished thirty-seven thousand units of housing annually. Decades later, the U.S. Department of Transportation would estimate that more than a million people were ultimately forced from their homes for highway construction. The people who lived in the neighborhoods labeled slums by white city planners had no way to fight back; the authority of eminent domain was absolute. By the time the Voting Rights Act passed in 1965, most urban highways across the United States had already been planned and built.

When I-10 was constructed through the Fifth Ward, the highway department closed the exit from I-59 that shuttled cars onto Lyons

Avenue and through the neighborhood's commercial district. Without a steady flow of customers, many businesses withered and closed. Nearly seven hundred families left the neighborhood; more than a thousand children scattered around the city. Without kids to play sports, Finnigan Park, where families had congregated to watch baseball games on sweltering summer nights, sat vacant. Slowly, the once-thriving neighborhood grew quieter, the streets emptier. Many whose homes were spared eventually left, seeking better jobs or schools elsewhere. "The streets were taken, the houses were taken, the people were taken. The familiarity of where you were raised was gone," O'Nari says. Their history had been erased, the community split and scattered. "It was like, go you must," she says. "No choice."

5

REMOVE

Dallas
I-345

The first highway in Dallas opened on a hot evening in August 1949, the hazy sky mottled with a few scattered clouds. Seven thousand people attended the dedication at the Ross Avenue overpass, just east of downtown. The celebration featured a "huge square dance for whites south of the Hall Street overpass, and an old-fashioned street dance for Negroes north of Hall Street," as *The Dallas Morning News* reported. Before the first cars rolled onto the expressway, the mayor's wife hurled a bottle of cologne onto a model of the highway, thus christening it.

In 1950, there were nearly fifty million vehicles on the road in the United States. A decade later, there were seventy-four million. There simply wasn't space for them all. As city streets became increasingly crowded, people clamored for something to be done. In Dallas, the city convened a citizens' traffic commission in 1955, which produced a film titled *Report to Dallas*. Dallas was approaching a million residents. Thousands of people moved to the city every month. According to an all-knowing narrator, "Dynamic growth brings people and growing payrolls. As business grows, population grows, and the snowball keeps rolling until there appears to be no limit to the area's potential." The camera pans to a freshly paved highway flowing with cars. "But there is a limit, and it's already pinching: The problem of traffic. A problem of Dallas progress."

This message was repeated in cities across the country: Traffic jams were strangling cities, impeding progress and economic development. Streets were choked with cars, and the only solution was more highways. In Dallas, the message was not subtle: "We must remember that traffic is the lifeblood of the city. Slow down the traffic and the heartbeat slows down. Stop it and the city dies."

North Dallas, where the Central Expressway began, was originally settled by a group of freed slaves, who purchased an acre of land on what was then the outskirts of Dallas for $25 and established a thriving community that grew to include thousands of residents. There were churches and elementary schools, Black doctors and dentists, grocers and movie theaters. When the Central Expressway was built, it tore through the center of the former Freedman's Town. "Questions are coming fast and furious from the many long time residents who must move as their homes must be torn down to make way for the Central Boulevard expressway," reported *The Dallas Express*, a Black-owned newspaper, on October 5, 1946. "The city has stated that October 15 is the final date for occupancy for a section that houses 1,500 persons. With winter just a few weeks away, their eviction notices just a few days away, no place available for them to rent and no housing construction for Negroes underway, these persons are facing a crucial time." Many people had lived in their homes their entire lives. "Every time the Negro gets a place that is decent, the white man finds some use for it," Ms. Anna Herdon, who lived on North Washington Avenue, told the paper. "They could not find anything else to do with it so they decided to make a boulevard of it."

As the North Central Expressway reached completion, the Texas Highway Department turned its attention south. At the time, South Dallas was a predominantly white neighborhood, full of middle-class Jewish families who lived in modest homes on tree-lined streets. In 1949, a real estate developer named Karl Hoblitzelle opened a grand theater on Forest Avenue, a major east-west thoroughfare lined with homes and businesses. The theater had a modern spun-glass screen for motion pictures, fifteen hundred plush push-back seats, and a stage for live performances. The Forest Theater would be the largest suburban theater in the Southwest—for whites only—and the first catering to families, with an innovative soundproof "cry room" for children. More than five thousand people attended the grand opening, where executives dedicated the theater to the "community spirit that has long placed the people of South Dallas among the pioneer leaders of Greater Dallas."

When construction finally began on the South Central Expressway, the highway narrowly avoided the new theater. In 1954, workers started pulling up wood railroad ties by hand from the abandoned Houston & Texas Central railroad tracks running along Central Avenue. The following year, an aerial photograph showed a swath of bulldozed land cutting south from Forest Avenue, as if a massive pencil eraser had been dragged through an otherwise tightly drawn neighborhood, packed with homes and trees. The trench stretched across a full city block, the houses on either side of it suddenly exposed, porches looking out over vacant space. In the bottom right of the frame, two car lengths from a giant berm of pushed dirt, rose the Forest Theater, barely spared. The marquee displayed showtimes for Lady and the Tramp.

More than thirteen hundred homes in South Dallas were demolished to make way for the South Central Expressway. Those that remained were gradually abandoned as the people who might have once been eager for the progress promised by the highway were confronted with its paved-over, hollowed-out reality. Black families moved into these abandoned homes, driven south by segregation and racial violence. Many had been displaced by the North Central Expressway a few years earlier. By 1956, when the final stretch of the nineteen-mile Central Expressway opened to traffic, South Dallas had changed so drastically that the Forest Theater briefly closed its doors and reopened as a "de luxe movie house for Negroes."

From the beginning, the plan had always been to connect the northern and southern segments of the Central Expressway with an elevated highway that ran over the eastern edge of downtown. There was already a six-lane road that connected the two segments, but traffic snarled at several light-controlled intersections. An elevated expressway would ease this congestion and help drivers get in and out of downtown more quickly. But like almost every urban highway built in America, the Central Expressway was running over budget; even the people who planned these highways couldn't conceive of how expensive they would be to build. In 1959, the Dallas Chamber of Commerce wanted to know what had happened to the plans for the elevated expressway. City and state highway officials had little to say on the record. "Privately, however, they have explained that they want to get [the project] made part of the federal interstate system, for which the Federal Government pays 90 percent of the costs and the state the other 10 percent," reported *The Dallas Morning News*. If they succeeded, Dallas could get its elevated highway built without paying another cent. In 1964—four years after Bragdon condemned precisely this kind of urban highway—the U.S. Bureau of Public Roads agreed to include the elevated expressway in the interstate highway network, shouldering 90 percent of the estimated $19 million required to build the 1.5-mile structure. Four years later, *The Dallas Morning News* noted that the connection would be known as Interstate 345. Before the elevated highway opened in 1974, a highway department engineer promised that it would become "one of the most imposing structures in Texas and probably the South. It will be a real landmark for Dallas."

But if anyone had listened to Bragdon, it never would have been built.

In 2002, when Patrick Kennedy loaded up his Toyota Corolla and moved to Dallas, sight unseen, I-345 was falling apart. Hundreds of fatigue cracks had started to appear in the floor beams, which held the highway up, and although TxDOT would spend millions to retrofit the aging bridge, the cracks kept coming.

Patrick had recently graduated with a degree in landscape architecture from Pennsylvania State University, where he'd become obsessed with urban planning. He didn't know much about Dallas, but he wanted to live somewhere with warm weather, and the city seemed like a place where he could tilt at windmills. "It was a chance to have a career's worth of work," he says. When Patrick moved to Dallas, he rented an apartment in Deep Ellum, a neighborhood just east of downtown, and walked to work along Pacific Avenue. Eventually, he got rid of his Toyota Corolla and started walking, biking, and using public transportation to get around the sprawling city. Every day, he'd walk on fractured sidewalks, past public storage facilities, boarded-up buildings, and surface parking lots, the black asphalt radiating heat in the summer. He'd walk under I-345, the cement structure casting a long shadow blocks beyond its physical reach. One day he realized: It doesn't have to be this way.

In 2011, the City of Dallas published a master plan for downtown that concluded that the inner highway loop was "a significant barrier to surrounding neighborhoods" but that fixing it was a "long-term and expensive proposition." By then, TxDOT had announced plans to rebuild every other highway that circled downtown, save for I-345. "I said, well, that's not right," Patrick recalls. He talked to a friend, a real estate developer named Brandon Hancock. For two years, they studied the traffic flows on the eastern edge of the city and saw that while the highway was often congested during peak hours, the city streets that surrounded it were largely empty. The highway had scrambled the local grid, disrupting the north-south flow of streets like Good Latimer and Cesar Chavez, which were built to carry much more traffic than they did. "We said, okay, hypothetically, what if we just remove the thing?" Patrick says.

Patrick and Brandon mapped the highway's right-of-way and the underdeveloped land that surrounded it and found that it affected hundreds of acres representing billions of dollars in developable land

and millions in property tax revenue, in a city that desperately needed the money to fund basic services like schools and sewage. North Texas was booming, but Dallas decidedly was not: Between 2000 and 2010, while the Dallas–Fort Worth metroplex grew by almost 25 percent, the city of Dallas grew by 0.8 percent, adding a grand total of 9,236 people. Job growth downtown had stagnated starting in the 1950s, even as the city continued to build high-rise office towers. All the jobs went north, along with the highways. By 2010, almost a third of Dallas's downtown office buildings and high-rises sat vacant.

Instead of spending hundreds of millions of dollars to maintain the structure, Patrick argued, TxDOT should just tear it down. "Here's a dirty little secret: you cannot cure congestion without killing a city, and no one congests places not worth visiting," he wrote in *D Magazine,* a monthly lifestyle magazine. No one wanted to visit downtown Dallas anymore, not since the highways had done what they were designed to do: draw people outward. Maybe a little congestion would do Dallas some good, particularly if it meant walkable streets that were full of things worth walking to.

In 2012, TxDOT held a public meeting about I-345. "They showed up with nine options," Patrick says—nine different ways to rebuild the highway. Patrick went to the meeting and said, "These aren't options, these are price tags." A columnist with *The Dallas Morning News* named Robert Wilonsky decided to write about Patrick's campaign to tear down I-345. "Kennedy's talked to the land owners, done the math and figures tearing out the highway will reconnect downtown with East Dallas, revitalize what's now 245 acres of vacant, wasted land . . . reduce traffic bypassing downtown, make better use of the surrounding streets on the grid, bring in some 20,000 new downtown residents and result in 'new walkable urban housing and high quality public space,'" Wilonsky wrote.

"That's when Victor Vandergriff emailed me out of the blue and said, 'What are you doing?'" Patrick recalls. Victor was a member of the Texas Transportation Commission, but he was unlike most of the power brokers he served alongside—curious and free from the dogma that dominated transportation planning. It was pouring rain when the two men met up under I-345, water streaming off the elevated structure. Patrick made his pitch. Victor was intrigued. He commissioned a study called CityMAP, published in 2016, which began with a question: What makes a great city? Rather than looking at one highway project in isolation, TxDOT considered the entire urban core and how the highway system connected people and places. CityMAP was unlike any document TxDOT had ever produced, representing an inflection point in how some TxDOT engineers, at least in Dallas, considered their work. "We're not here to move cars," says Mo Bur, TxDOT's Dallas district engineer. "We're here to have a project that will address all kinds of transportation. And then, once we leave, what does it do to the community that the highway went through?"

For the first time in the agency's history, TxDOT studied the full removal of an elevated interstate highway. TxDOT engineers talked to hundreds of people across the city. Public feedback was decisive: People didn't care all that much about congestion. They cared a lot about community and connectivity and parks and housing. Removing I-345, the study found, would increase congestion delays on nearby thoroughfares by just one minute.

The idea of highway removal has been around as long as highways themselves. In the mid-1950s, the California Highway Commission started construction on Interstate 480, known as the Embarcadero Freeway, to connect the Bay Bridge on the east side of San Francisco with the Golden Gate Bridge. More than a mile was built, obstructing views of the bay and cutting off Telegraph Hill from the waterfront. In 1959, after thousands of people protested, San Francisco's board of supervisors adopted a resolution opposing further construction not only of the Embarcadero but of all freeways in the San Francisco Master Plan, citing "the demolition of homes, the destruction of residential areas, [and] the forced uprooting and relocation of individuals, families and business enterprises." In 1986, city leaders proposed tearing down the highway. In a contentious local election, voters rejected the proposal, convinced that removal would cause gridlock. San Francisco's mayor, Dianne Feinstein, called the defeat "an anti-environmental vote . . . a victory for the automobile."

On October 17, 1989, a twenty-five-mile section of the San Andreas Fault ruptured, sending a 6.9-magnitude earthquake across the bay. Earlier that afternoon, a woman named Margaret Thomas left the hospital where she worked as a maternity nurse. After she collected her youngest daughter, Teresa, from her mother, who had been watching the baby, Margaret stopped at a nearby grocery store on her way home. At 5:04 P.M., as she was checking out, the store began to shake. Margaret looked at the cashier. She looked at Teresa, sitting in the cart. And then she looked back at the cashier. The cashier was gone. So Margaret grabbed Teresa and ran outside to the parking lot. As soon as the ground stopped shaking, she ran back inside for her groceries—she'd already paid for them—strapped Teresa into her car seat, and accelerated onto the freeway on-ramp on Fell Street, rushing home to her two older daughters. The next morning, she saw on the news that the on-ramp she'd used to get on the highway had been condemned, so damaged by the shaking it began to crumble.

Margaret had been lucky. Sixty-three people died in the earthquake, most of them on or near the Bay Area's elevated freeways. A stretch of Interstate 880 in Oakland collapsed, killing forty-two people. The Embarcadero and Central Freeways had so much structural damage that they were both closed to traffic.

In the aftermath of the earthquake, when it was revealed that fixing the Embarcadero would cost as much as rebuilding it, the newly elected mayor, Art Agnos, argued that the city shouldn't squander "the opportunity of a lifetime." While city leaders debated what to do, drivers maneuvered around the highway and got where they were going without noticeable delays. People drove less. Transit ridership increased 15 percent. Public opinion turned: Maybe the Embarcadero wasn't so necessary after all. In 1990, the board of supervisors voted 6–5 to demolish the broken structure. Demolition began in 1991. A decade later, the elevated structure was transformed into a populated, palm-tree-lined boulevard, soon to be surrounded by apartments and restaurants and offices.

Just as the Texas Highway Department was beginning construction on I-345, across the country, the nation's first elevated highway was being torn down. Erected in the 1930s under the gaze of Robert Moses, the West Side Highway was once a "gleaming new concrete ribbon" that spanned the western edge of Manhattan. Four decades later, the highway was decaying. One day in 1973, a truck carrying sixty thousand pounds of asphalt on its way to repair the elevated highway instead fell through it, when a portion of the bridge collapsed under the truck's weight. Just as in San Francisco, while officials closed the highway and debated about how to rebuild it, people made other choices. Traffic volumes on nearby streets decreased by 53 percent. After a prolonged, decades-long debate, in the 1990s the city demolished the structure and built a boulevard in its place.

The first American highway to be intentionally removed—not by earthquake or accident—was Portland's Harbor Drive, which was built in 1950 and ran north-south along the western bank of the Willamette River. In the late 1960s, a group in Portland called Riverfront for People started campaigning for the removal of Harbor Drive. By then, Interstate 5 ran along the eastern bank of the Willamette, rendering Harbor Drive redundant. Riverfront for People had a sympathetic ear in their newly elected governor, Tom McCall, an environmentalist, who appointed a task force to study the effects of tearing down the highway. In 1974, the state closed the three-mile stretch of highway and began constructing a thirty-seven-acre riverfront park where cars had once streamed.

In 2004, Milwaukee's mayor, John Norquist, who had successfully campaigned to remove a mile-long spur of the elevated Park East Freeway in his hometown, stepped down to lead a nonprofit called Congress for the New Urbanism, where he began a campaign for the removal of urban highways across the country. By the time Patrick made removing I-345 a citywide talking point in Dallas, Congress for the New Urbanism had active campaigns in two dozen cities, from Pasadena, California, to Syracuse, New York. Some campaigns called for full demolition, replacing highways with boulevards as in San Francisco; others proposed what's called a cap and stitch—depressing a highway and covering it with a park or promenade at street level, as Dallas did over the Woodall Rodgers Freeway in 2012. To date, eighteen North American cities have either replaced or committed to replace a limited-access highway with an urban street. In 1998, a global study of more than a hundred projects that reduced road capacity found an average traffic reduction of 25 percent, "even after controlling for possible increased travel on parallel routes."

Most campaigns focused on the economic benefit of highway removal—all that land that could be put to more productive use. In Milwaukee, for example, the removal of the Park East Freeway cost $25 million and generated more than $1 billion in new private investment around the former highway. In Dallas, a 2021 study from Toole Design Group found that 377 acres of land in the core of the ninth-most-populous city in the country had been either consumed by I-345 or depressed by its presence. This represented more than $9 billion in developable land and $255 million in annual property tax revenue for Dallas. Removing the highway could create the capacity for fifty-nine thousand new jobs downtown and twenty-six thousand housing units, the study found.

Three decades after the Loma Prieta earthquake, Margaret Thomas moved to New Orleans, where her mom had grown up and she'd spent her childhood summers. Her extended family lived in the Lafitte Projects, a public housing complex located on Claiborne Avenue. Before the Claiborne Expressway tore through the Tremé and Seventh Ward—some of the earliest Black settlements in the country—hundreds of sprawling live oaks lined the wide boulevard, the heart of New Orleans's Black business district. People called this shady space between the two sides of the street "the neutral ground."

In 1966, on Ash Wednesday, the bulldozers came. They took the trees, along with hundreds of businesses and homes. Margaret remembers how upset her family was when the expressway came through, the violence of the demolition. "Their shopping was being taken away; their community was being disrupted," she says.

In 2012, Margaret and her husband bought a two-story building in the shadow of the expressway. The building—a former bank that served the businesses on Claiborne—had sat vacant for several decades by the time Margaret bought it. It had high, arched windows and white stucco walls. It needed work, but it was unquestionably grand. They set about transforming the space into an event venue. Margaret hadn't thought much about the Claiborne Expressway until she started spending every day in its shadow. All day long, she listened to the rattle and thrum of the interstate and breathed in its polluted air. The grit from the highway stained her white walls, which required constant cleaning. As Margaret and her husband were getting to know the community, she met a woman named Amy Stelly, an architect who had started a campaign to remove the Claiborne Expressway. Amy grew up two blocks from the elevated freeway and, after leaving New Orleans and raising kids, returned to live in the house she grew up in. "I've been committed to seeing that highway removed since I was a kid," Amy says. "Intuitively, I knew it was harmful and decided this is something I'm going to do when I grow up and I know enough."

Margaret joined Amy's campaign almost immediately. "The highway should have never run through a neighborhood," she says. She knew removal was unlikely, but she also knew it was possible. She'd seen it happen. Before the earthquake, the Embarcadero carried "an amazing amount of traffic," she says—more than 100,000 cars a day. So did the Central Freeway and Oakland's Interstate 880. After the earthquake, "The traffic got rerouted, went on different streets, it was fine," she says. "I saw it happen, all three of those freeways. None of those cities crumbled and fell into the ocean. Traffic just went a different way."

In 2013, Patrick helped launch a political action committee called Coalition for a New Dallas that supported the removal of I-345. For Patrick, tearing down the highway required a radical reconsideration of how to accommodate and plan for growth in sprawling Dallas. The highway had "eliminated the value of proximity" by making it cheap and easy to spread out, he says. "The fifties, sixties, seventies was this big bang that separated the city. How do we come back together?"

For almost a century, Elm Street was the heart of the neighborhood just east of downtown known as Deep Elm, a name eventually transformed by syrupy southern speech into Deep Ellum. Originally settled by former slaves after the Civil War, the neighborhood became the cultural and commercial heart of Black Dallas. Businesses like La Conga Cafe and the Harlem Theater offered live jazz, gospel, blues, and vaudeville shows. A photograph from around 1937 shows a throng of people, white and Black, walking on a sidewalk on Central Avenue south toward Elm Street. Above them hangs a sign for the Gypsy Tea Room, a club owned by a Black musician originally from Louisiana. Dozens of theaters and clubs along Elm Street would launch the careers of jazz and blues musicians in the 1920s and 1930s. As European immigrants—many of them Jewish—settled in the neighborhood and opened businesses, Deep Ellum became one of the city's only integrated communities. By the 1940s, Deep Ellum was "the gathering place of Blacks from all over the country," wrote the historian Michael Phillips in his book *White Metropolis*.

A Polish immigrant named Harry Wilonsky owned an auto repair shop called S&W Auto Parts on Elm Street. Harry would often leave the shop for lunch, passing the sprawling chaos of Honest Joe's Pawn Shop and the majestic Knights of Pythias Temple next door. The Knights of Pythias Temple was the first major commercial structure in Dallas built by and for Black businesses. Designed by William Sidney Pittman, the first licensed Black architect in Texas, it housed professionals like lawyers, dentists, and doctors. The streets of Deep Ellum were usually packed with people. "He would have to push people out of the way just to get back to the store," says Robert Wilonsky, Harry's grandson and the journalist who first wrote about Patrick's campaign to remove I-345. "He very vividly remembered Deep Ellum being very Black and very white and very mixed and very loud. He didn't live in New York, but he always thought that where he was in Dallas was as close as he would ever get." In 1955, after Central Expressway came through Deep Ellum, Harry moved his auto shop to South Dallas, on the other side of I-30, following his customers as they too flowed south. "There was an exodus from downtown when Central began," Wilonsky says. That was the whole point of the highway, after all—to make it easier for people to leave.

Construction of I-345 began in 1968. The highway consumed the 2400 block of Elm Street, the heart of Deep Ellum's commercial district. In 1950, twenty-nine businesses occupied the block, starting with Oatis Drug Shop and ending, on the east end, with Rose Dress Shop. In between, there were stores for photographs and sandwiches and jewelry and haircuts. There were pawnshops and liquor stores and a sportsman's club for billiards. There were repair shops—for cars and trucks, shoes and windows. There were restaurants and cafés and nightclubs. By 1965, the block had begun to empty out. Eleven addresses were listed as vacant in the city directory that year, while only sixteen businesses remained. By 1975—the year after the highway opened—addresses on that block of Elm Street were no longer listed as vacant. They simply no longer existed. The city directory skips the 2400 block of Elm altogether, jumping from "2301 Elm Street, Home Furniture Co." straight to "2511 Goodwill Industries of Dallas."

And that was just one block. The highway consumed a total of fifty-four city blocks. To build I-345, the Texas Highway Department purchased 133 properties comprising just over twenty acres of right-of-way. Only forty-one property owners contested the price they were offered by the state. The total cost just for the land was $14.7 million, roughly $130 million in today's dollars.

Dallas was a city "hostile to remembrance of history," Phillips said. Like that of most American cities, Dallas's history is rooted in racism. But the accommodation that crafted Dallas was a tacit agreement not to acknowledge that racism. "In this obsessively image-conscious city, elites feared that a conflict-marred past filled with class and racial strife represented a dangerous model for the future," Phillips wrote in *White Metropolis*. "City leaders transformed the community into a laboratory of forgetfulness."

Part of how they did that was to etch this forgetfulness onto the landscape. To pave over Black prosperity so that fifty years later they could say, There has never been any other way.

6

EXPAND

Houston
I-45

In 2019, in the same month Modesti Cooper opened the letter from TxDOT and found out she might lose her brand-new home, hundreds of people crowded into a conference room on the first floor of a towering black-glass building on Houston's east side. It was hot and clear, the bright blue sky reflected in the building's dark exterior as people filtered inside the ground-floor conference room, wrapping themselves in cardigans and blazers against the aggressively air-conditioned interior. There weren't enough seats for everyone who had come, so those who couldn't find a space standing against the back wall were shuttled into an overflow room.

The meetings of the Houston-Galveston Area Council's transportation policy council were usually subdued affairs. Few people knew the governmental body existed, much less what it did. Enabled by the Federal-Aid Highway Act in 1962, metropolitan planning organizations were regional governments, tasked with coordinating across cities and allocating federal transportation funding. The bodies consist of elected officials—mayors, city council members, county judges—who serve alongside staff planners and engineers. Democracy, but at a remove.

The members of the Houston-Galveston Area Council's transportation policy council had convened to consider allocating $100 million to a $7.5 billion project to rebuild and expand nearly twenty-five miles of highway that threaded through the center of Houston. The North Houston Highway Improvement Project would be the largest road project in Houston's history. It would expand Interstate 45, which swept through downtown Houston and stretched north to the suburb of Spring, but it would also rebuild and reroute Houston's downtown loop, a tangle of lanes and overpasses that included Interstates 10 and 69. The project had been in the works for nearly two decades, cumbersome and slow but constantly justified by Houston's incredible growth. Since 2000, Houston's population had doubled, and the boom showed no signs of stopping. As always, there were the traffic models—without added capacity, travel times would increase by nearly 40 percent, TxDOT said—and the safety justifications. I-45 was the most dangerous stretch of highway in the state, and the agency that had built it that way insisted it was the only one that could fix it.

In 2017, TxDOT released a draft environmental impact statement for the project, documenting for the first time the agency's specific plans for the expansion—where it would go and whom it would affect. The document stunned Houstonians. The footprint of the expanded highway was enormous. For most of its length, I-45 would expand by 50 percent, spanning 430 feet across. Through downtown, I-69 would grow from 220 feet across to 570 feet—a tenth of a mile wide. The increased traffic accommodated by the expanded highway would increase air pollution in neighborhoods that already had higher than average rates of childhood asthma. All along, TxDOT had said the project was likely to displace some properties and homeowners. But the scope of that displacement was staggering. Under TxDOT's recommended plan, more than twelve hundred residences would be demolished, including two public housing complexes and hundreds of single-family homes and apartments. More than 300 businesses employing nearly twenty-five thousand people would be displaced. So would four places of worship and two schools. TxDOT's own analysis concluded that "all alternatives would cause disproportionately high and adverse impacts to minority or low-income populations."

For years, transportation advocates and city planners had been trying to work with TxDOT to change the scope of the project. Many agreed I-45 should be rebuilt but wanted to keep the road within its current footprint and instead create new capacity for buses and light-rail. But when the draft environmental impact statement was published, it became clear that TxDOT would do what TxDOT wanted to do, and what TxDOT wanted to do was make space to move cars.

Shortly after the initial plans were published, a retired nurse named Susan Graham went to a community engagement meeting in her neighborhood, the Near Northside, a quiet, historically Hispanic community just east of I-45. Before she retired, Susan ran clinical trials within the University of Texas hospital system in Galveston and later Houston, focusing on psychiatric care, addiction, and psychiatric treatment. At the University of Texas Health Science Center, she ran clinical trials for the neurosurgery department. She retired at sixty and moved to the Near Northside. "It just looked like a place where you would have community," she says. When she heard about the I-45 expansion, she thought, *no way.* A giant freeway contradicted everything the city's elected officials said they valued— walkability, connectivity, climate action. At that community meeting, hosted by TxDOT, she didn't feel very engaged. "We were basically just told what TxDOT was going to do." After the meeting, Susan vented to a neighbor who'd gone with her. The more they talked, the more lathered up they got. They decided to do something to register their opposition to a project that the state was selling as a done deal. "What about yard signs?" Susan wondered.

"What should they say?" her neighbor asked.

"How about 'Stop TxDOT'?" Susan said.

They threw "I-45" on the end for good measure and designed a simple graphic in stop-sign red with white sans serif lettering. They raised some money on GoFundMe to print the signs and started handing them out along with flyers about why they didn't think the project would be good for their neighborhood or the city. That summer, a local community development corporation offered them meeting space. More than sixty people showed up. "We were absolutely blown away," Susan says. "We knew that there were a lot of people that were very, very concerned about this project, but they felt like they had absolutely no voice."

A month after the yard signs had sprouted on lawns, Susan steps up to the lectern to speak to the two dozen members of the transportation policy council. These officials are not used to hearing hours of public testimony. The layout of the conference room is unintentionally revealing, with four of the members sitting with their backs to the audience, facing the interior of the square-shaped conference table. "Both TxDOT and the City of Houston have held the obligatory community outreach meetings," Susan says. "But have we felt heard? No, I don't think so." Behind her, the room is packed, every chair occupied, every inch of wall space covered. Most people wear oversized red stickers tacked onto shirts and dresses: DELAY THE VOTE. With few exceptions, the sixty-two people who have shown up on this sweltering July day to testify oppose the I-45 project. One after another, they insist that widening the highway will increase air pollution, worsen flooding, displace families, and destroy jobs—and it won't even fix traffic. The hours drag on. The chairman of the transportation policy council interrupts to announce that they have ordered pizza and it will be delivered shortly to the overflow room.

A woman named Molly Cook approaches the lectern, also wearing one of the giant red stickers over a maroon sheath dress. Thick red hair frames her face as she introduces herself as an emergency room nurse at the Texas Medical Center. She tells the council that she grew up in Spring—the booming, subdivided suburb on the northern end of the I-45 expansion. "I have a very vested interest in improving efficiency on I-45. I grew up on it," she says.

Every Wednesday evening, Molly's mom, Judy, would drive her into

Houston on I-45 for harp lessons. They would talk or listen to music, and she'd gaze out her window at the blurred city streaming past. I-45 was her family's tether to Houston; her dad, Mark, commuted to work on the highway every day for a decade. It provided access to entertainment—Mark and Judy had season tickets to University of Houston football games—and medical care at Houston's world-class hospitals. The freeway had been built for her, for families like hers— white families, mostly, who bought expansive homes in the sprawling suburbs with the promise of speedy access back downtown.

But Molly does not want this highway to be expanded. "I do not support the expansion of Interstate 45 and certainly not the plans as they stand today," she says. "The project is going to uproot families, tear apart communities, and destroy thousands of jobs. We know that change and progress come at a cost. But this is neither change nor progress. This is a repetition of our past mistakes."

After four hours of public testimony, the council discusses the funding allocation. The Houston-Galveston Area Council does not have much money to give—only $100 million of a $7.5 billion project—but the vote matters. State and federal funding comes with few strings attached, but one of those strings is a show of local support. Lina Hidalgo, who had been elected as the chief executive of the county the year before, when she was only twenty-eight, insists that they need more time to study such a massive, consequential project. "This body is being asked to provide seed funding, essentially, for a project . . . that as it stands is deeply problematic," she says. "If this feels wrong and it feels rushed, it's because it is wrong and it is rushed." Her opposition will prove fruitless in this moment, but it is an early sign that this project might not proceed like every other highway project in TxDOT's arsenal. Her colleagues insist it's now or never. There's some sense that the dollars will evaporate if the body delays their vote of approval. "If we delay . . . we send a message that says, hey, we're not ready," says one council member. "If we don't show the transportation commission that we have an interest in this project, these dollars will be gone." Despite the hours of testimony, the organized opposition, the red stickers, and Hidalgo's plea, inertia is on TxDOT's side. The council approves the $100 million funding request.

Back outside in the glaring midday sun, Susan approached Molly and told her about the yard signs. "She was like, girl, you have got to come hang out with us," Molly says. "I was like, whatever, I don't know who this lady is." The following month, at yet another TxDOT community engagement meeting, Molly ran into Susan again. Susan was insistent: Molly should come check out their fledgling opposition group. Molly liked Susan's persistence, so one night a few weeks later, she went to a meeting and met a dozen other people who had heard about the group named after the yard signs: Stop TxDOT I-45. Molly asked, Has anybody made this an election issue? Why aren't we making city council candidates talk about this? Nearly $8 billion was on the line, a once-in-a-generation infrastructure project that would affect every Houston resident, directly or indirectly. Susan said, "Well, do you want to chair the election committee?"

Molly started organizing block walks in the neighborhoods adjacent to the highways that would be expanded. Some people had attended a community meeting five years earlier, but they hadn't heard anything since then, so they assumed the project had been shelved or passed over. Most people weren't aware of the expansion. People living on land that TxDOT wanted to take for the highway didn't know they were going to lose their homes. Either they were renters and hadn't been told, or they'd missed the letter, shuffled in among advertisements and solicitations.

One weekend in December, Molly organized a canvass in the Fifth Ward, just south of I-10. The volunteers dispersed, armed with clipboards and a newsletter sign-up sheet. A guy in his early thirties named Neal Ehardt walked up Nance Street, which runs alongside I-10, and tried to enter the tall, angular house on the corner.

As she settled back into life in Houston, Modesti had kept busy, exploring all that had changed in the city since she left. She walked to Discovery Green, a twelve-acre park in downtown Houston, for live music. She joined a gym nearby. She rode her bike on the new trail that threaded alongside Buffalo Bayou. She started a catering business, posting weekly menus on her Instagram account and delivering food across the city. Her house quickly became the gathering grounds for her family, who lived dispersed across Houston. Her godchildren swam in her pool and played video games in her upstairs den. Her grandmother lived fifteen minutes away, and although she'd stopped driving on the highway, she could get to Modesti's house using only neighborhood streets.

Modesti had installed security cameras covering every edge of her property. She was guarded, suspicious of strangers. When she caught a glimpse of Neal trying to find a way onto her property, she walked out on her second-story balcony and called down to him, "Who are you looking for?" Neal responded that he was a volunteer with a group called Stop TxDOT I-45. "We're trying to find everyone who's affected by the highway expansion. Do you know anything about it?"

Modesti started going to Stop TxDOT I-45's meetings. She had no idea how many people were affected by the highway expansion. "I didn't know there were more people going through the same situation," she says. "I didn't even know there were groups that were advocating against it." Many of the people who showed up at Stop TxDOT I-45's meetings lived at a safe distance from this particular highway expansion, but they were outraged on her behalf. They had long lived with the effects of Houston's auto dependency. They had asthma. They'd lost loved ones to car crashes. They'd had close calls biking around the city. They thought Houston's addiction to automobiles had made their home a hotter and more dangerous place to live. "I didn't know I had help," Modesti says. "I thought that I was going to have to listen to TxDOT and just turn over my home. Stop TxDOT gathered everyone and said, 'We're going to fight for you guys.'"

In 1981, Mark and Judy Cook were in their early twenties and newly married. They'd recently moved to Houston from Jacksonville, a small East Texas town three hours north of Houston. Judy had been hired by NASA while she was still in college, so she finished her degree at College of the Mainland, zipping back and forth along I-45 from Clear Lake to Texas City. Mark didn't know what he wanted to do, but after bouncing around odd jobs, he decided he'd better get a college degree, too, and graduated with a bachelor's in accounting from the University of Houston. One day, they got a flyer in the mail for a new subdivision selling homes in northwest Houston. The developer, eager for buyers, promised a free TV to anyone who toured the model home. Mark and Judy needed a TV. So they drove up Interstate 45 to Farm to Market Road 1960. As they traveled north, the development along the highway dwindled, the homes and gas stations and feeder road strip malls receding into lush forest and open fields dotted with grazing cattle. Somebody had thought to buy one of those fields and plat the land for single-family homes. "We thought, who in the world would ever live this far out?" Mark remembers. "That's stupid." They wandered through the model home, didn't get the free TV they were promised, and drove back to the city, marveling at the distance some people would drive for a big backyard.

A decade later, Mark and Judy became those people. By then, they had two kids, a son named Kevin and a daughter named Molly. They liked North Houston, which unrolled into the Piney Woods region of East Texas, with lush stands of loblolly pine and sugar maple and red cedar. "We're tree people," Mark says. In 1992, Mark and Judy bought a home in Spring, five miles north of the farm to market road that had once seemed like the end of the earth. But by then, Houston's urban footprint had poured over the fields of grazing cattle, big-box stores sprouting like mushrooms as hardwoods were felled and the ground below them was scraped clean.

As the Cook family settled into their new home in Spring, Mark commuted downtown to his job at Enron. When he was hired, the company was called Houston Natural Gas, but it had since been acquired by a company called InterNorth, and after a brief stint as

Enteron—which turned out to be the Latin word for intestine, leading to the joke "We'll pass your gas!"—the company became known, infamously, as Enron. In those days, a decade before scandal and bankruptcy, Enron was the place to work in Houston. The company subsidized bus fare, so for nearly a year after they moved, Mark took the bus downtown. "The buses were heaven on earth," he says. Every morning before dawn, he'd shuffle out the sports and business sections from the newspaper and leave the rest for Judy to read. He loved the bus—the time to read and sleep and sit quietly. But as his responsibilities at work grew, his hours became increasingly unpredictable. A few times, he missed the last bus back home and got stranded downtown. "I said, I can't do that anymore."

He started driving, an hour there and an hour back. Mark and Judy had three kids by then, and Mark often left before they woke up and arrived home after they'd finished their homework, dinner on the table, the day nearly done. When she was little, Molly sometimes slept curled up on the stairs' carpeted landing. "I wanted to see him before he went to work. Because I didn't see him," she says. Sleepy, she would stand up to give him a hug and then crawl back into bed.

Eventually, Mark quit Enron and started a natural gas storage company with three friends. It was years before the fracking revolution would transform West Texas, but there was still plenty of natural gas being excavated across Texas. Mark's company drilled deep into salt domes scattered along the Gulf Coast and pumped water thousands of feet below the surface, washing out the salt to form giant underground caverns. "You can use a compressor to push natural gas in there, and it holds it like a thermos bottle," Mark says. When natural gas was cheap and plentiful, companies and utilities would store it underground, leasing space from Mark's company until there was a surge in demand.

Mark's new office was west of downtown, forty miles from their home. On good days, Mark could get to work in an hour and a half. But those were good days. Sometimes, he would sit in traffic for two hours, only to give up, turn around, and drive home. "I'd call my partners and say, 'Hey, I'm going home, I can't find a route to get me to the office. I'll work from home today.'" He could have worked from home every day; he was a salesman, either on the phone or on a plane traveling to meet a customer. But his partners wouldn't stand for it. There were only four of them running the company, and they wanted a "brain trust" in the office, he says. And besides, his partners commuted in every day—why shouldn't Mark? "They were livid if I didn't come in," he says. "It was peer pressure. I could have worked from anywhere."

By then, Mark and Judy had moved even farther north on I-45, to a master-planned community called The Woodlands. In 1964—the year after I-45 opened to traffic—a man named George P. Mitchell bought twenty-eight hundred acres of forested land thirty miles north of downtown Houston along I-45. Mitchell grew up in Galveston and graduated with a degree in petroleum engineering from Texas A&M. In the early 1940s, he settled in Houston and started drilling for oil. By the time he started buying up land along I-45, Mitchell had grown Mitchell Energy & Development Corporation into one of the country's largest producers of oil and natural gas. Back then, you could extract oil from deep within the earth and still consider yourself an environmentalist, as Mitchell did. He loved trees. He hated Houston's urban sprawl, how its bayous had been paved over and forests replaced by strip malls and suburbs.

"We're stuffing up the highways back and forth," Mitchell told *The Houston Post* in 1972. "We've got to urbanize, but how?" Mitchell's answer was the anti-suburb—a self-contained community, integrated with its natural environment, where families could access everything they needed for daily life, including jobs and schools. By 1974, he'd assembled nearly eighteen thousand acres of Piney Woods and announced his vision for "a new hometown" called The Woodlands. This new town would be populated by 150,000 people spread across seven distinct "villages," each wrapped in pine and oak trees and contoured by natural creeks. There would be jobs for 40,000 people. "No mere residential subdivision, The Woodlands will contain all services of a modern city," promised a brochure for the development. On Sunday, October 20, the *Houston Chronicle* ran a two-page spread announcing The Woodlands' grand opening, inviting Houstonians to drive twenty-five miles up I-45 to see "a whole new hometown." "Haven't you always wished that you could return to a relaxed, small town way of life? You would know your neighbors. Your children could walk down a forest path to school. You could get off the freeways and get back to living."

A decade later, this idyllic vision had yielded to a more persistent reality. "This forested hamlet twenty-eight miles north of Houston has the accoutrements of success," reported *The Houston Post* in 1985. "Its well-kept streets, lush lawns and modern buildings all speak of a prospering community. What it doesn't have yet is enough jobs." Half of the community's workers still commuted daily to jobs in Houston, the *Post* reported. And The Woodlands had become glaringly white. Nearly 20 percent of Harris County was Black, and yet only 1.5 percent of the twenty thousand people who had moved to The Woodlands were Black. A study conducted by Texas Southern University the year before found that "blacks were not inherently opposed to the community," but that they didn't move, because "they could not afford to live so far away from places where job opportunities existed."

By 2019, The Woodlands had become yet another suburb, "the same sort of exclusive retreat that whites chose in the 1960s as they fled the decaying inner cities—albeit better planned and with more appealing amenities," writes the journalist Loren Steffy in his book *George P. Mitchell: Fracking, Sustainability, and an Unorthodox Quest to Save the Planet.* "Designed as a remedy to urban sprawl, The Woodlands now accelerates it."

Molly Cook spent most of her adolescence riding her bike around the dendritic streets of The Woodlands. The neighborhood was full of kids and they pedaled along the quiet suburban streets that branched into dead-end circles. They walked to the shopping center down the street, McDonald's and Blockbuster hidden behind a curtain of pine trees. A creek ran behind their house, and Molly spent hours exploring the hidden wilds of the suburbs, traipsing over dry leaves and peering into puddles. The Woodlands was safe and it was quiet, but Molly never felt as if she belonged. "I was not very Abercrombie and everyone had Abercrombie," she says. She had untamed red hair, scraped knees. As she got older, Molly became increasingly troubled by the homogeneous affluence that surrounded her. Many of the people who lived in The Woodlands worked for Mitchell Energy. She didn't know the term "house poor" then, but that's what she saw—families living in half-a-million-dollar homes who couldn't afford to furnish them.

Every Cook kid was required to play an instrument, and harp was chosen for Molly. Her teacher lived near the Texas Medical Center, so every Wednesday afternoon Molly and Judy would climb in the car and accelerate onto I-45. "We're both talkers, so we would just chitchat the whole way down," Molly says. The movement liberated them, allowed them to talk about things that might have otherwise been hard to discuss. "I always liked having conversations with people driving because you're both facing forward," she says.

It was only when Molly left Texas and moved to Baltimore to attend graduate school at Johns Hopkins University that she realized freeways were not inevitable—an immutable part of the urban landscape. "I was shocked in public health school to understand that freeways were violently built through Black, immigrant, poor neighborhoods," she says. "It was something I'd never remotely considered." In Baltimore, she took the train everywhere she needed to go. It put her upbringing in a new perspective. "It was like, damn, my dad spent an hour and a half driving both ways on this horrible freeway," she says. By then, she'd been to cities like New York and Paris and loved walking to access everything she needed. She thought walkability "was something that happened somewhere else," she says. But in Baltimore, she realized car dominance was a choice— that freeways were a policy decision that enabled her life in The Woodlands, her dad's commute downtown, his job in oil and gas.

In 1996, an engineer working at Mitchell Energy discovered a new method for extracting natural gas from shale formations, known as fracking. Until then, most wells were fracked by injecting high-viscosity gels into shale, releasing natural gas. What if instead of the gel, which was prohibitively expensive, wells were injected with water and sand, which would hold open the fractures in the rock? Over the coming decades, this discovery would transform the global energy market as natural gas poured out of West Texas shale formations and the United States became the world's biggest producer of crude oil and natural gas. But before all that, Mark Cook would make a small fortune when he and his partners sold their company, having figured out a way to store all that natural gas back inside the earth.

After graduate school, Molly returned to Houston and volunteered for Beto O'Rourke's 2018 U.S. Senate campaign, knocking on thousands of doors across Harris and Montgomery counties. After Beto lost the election, Molly got a job as an ER nurse, but she had energy to burn. She decided to work on air quality. "Your lungs are the number one way that your inside world interacts with your outside world," she says. In 2019, she showed up at a breakfast hosted by a nonprofit called Air Alliance Houston and learned about a massive highway expansion that would bring more traffic—and therefore air pollution—to some of Houston's poorest neighborhoods. "I was just like, oh, damn, that doesn't seem like a good idea." When she learned about the I-45 expansion, she realized that this was her fight. "It's too on the nose," she says. "If I don't do this, whose job is it? I had a sense of, 'Daddy has all this oil money.' How do I reconcile that? How do I live a life protecting the environment while I literally inherit wealth from this other thing?" she says. "The freeway became a very concrete way to begin addressing those issues that seem so overwhelming you can't do anything about them. It's like, well, now you can. There's something to fight and work toward."

7

AGAIN

Austin
I-35

The neon-pink tape begins on the curb of the I-35 frontage road. It continues straight through the parking lot of Stars Cafe, under a metal railing, across the patio, up the white-brick exterior of the diner—past a banner announcing, BEER WINE & MIMOSAS HAPPY HOUR ALL DAY. Behind the building, the tape's trajectory resumes, traversing the white brick down to the black asphalt and into the alley behind the restaurant. The tape spans 107 feet—the amount of land TxDOT intends to take to widen I-35 through the heart of Austin.

Natasha Harper-Madison stands on the pink tape, behind a lectern affixed with the words HEAL THE SCAR OF I-35. Traffic roars on the upper deck of the highway, blanketing the gathering with an oppressive roar. From the lectern, Harper-Madison, the only Black member of Austin's city council, looks out past the fifty-odd people assembled for this press conference. Jay Blazek Crossley—the man who had called into the Texas Transportation Commission to oppose the funding for this expansion a year earlier—stands behind the speakers, wearing a navy suit coat, dress slacks, and gray slip-on Vans. He squints out from under a bright green trucker hat, the front stitched with the logo for Farm&City. Harper-Madison points to I-35, looming overhead. "This stretch of highway right here . . . is the physical manifestation of a segregationist past," she says. Harper-Madison grew up in East Austin when it was still a predominantly Black neighborhood, segregated by the highway. "It is the poster child of car-choked mobility systems. It pumps countless tons of carbon and pollutants every single year into the air we breathe. And right now, TxDOT thinks the best way to build I-35 is to build *more* I-35."

In August 2021, TxDOT published schematics showing three designs for an expanded I-35. At a public meeting at Huston-Tillotson University, a historically Black college in East Austin, engineers unrolled forty-foot-long printouts onto folding tables. Superimposed onto black-and-white aerial photographs of the city were bright yellow and orange and blue highway lanes, weaving one on top of the other, an intricately engineered braid. TxDOT intended to build a twenty-lane highway, including frontage roads. It would be a massive road, eight lanes wider than the twelve-lane highway that currently passed through the city. Stair-stepping around the highway, noted with a dotted blue line, was the new land TxDOT intended to take to expand Main Street Texas. The highway, as proposed, would consume dozens of acres of land and displace more than a hundred homes, businesses, and community centers.

When Tucker Ferguson, TxDOT's Austin district engineer, presented the state's plans to the Austin City Council the day before the press conference, Harper-Madison had asked him, given Austin's projected growth, how long it would take for the new I-35 to become just as congested as the existing highway. Ferguson had responded with surprising frankness: "We're not pretending to say that the expansion that we're proposing is building our way out of congestion."

Why was TxDOT planning to spend billions on a project that wouldn't even fix traffic? Harper-Madison asks now. "When you add more space for cars, more cars will always, always fill up that space. . . . We're talking about a $5 billion project that will cause a decade of displacement and disruption, and on day one it could be even more congested than it is right now." Harper-Madison is joined by three of her colleagues on the ten-member city council. One by one, they express their opposition to the project. Unfortunately, although the highway runs through the heart of Austin, today's opposition is largely rhetorical. TxDOT owns I-35 and TxDOT answers to the state, not the city. Still, Austin's mayor, Steve Adler, is conspicuously absent. Adler was a year from terming out of office, and he'd spent much of his time as mayor on defense as Governor Abbott stripped authority from Texas's liberal cities. Rumors had started to circulate that Kirk Watson was thinking about running to replace Adler. Watson had already been mayor of Austin, from 1997 to 2001, and had spent most of the last two decades representing the city in the state senate, where he served as the vice-chair of the senate transportation committee. In 2019, he'd filed a flurry of bills trying to get the funding allocated to expand I-35 in Austin. "There is no highway fairy, money doesn't grow on trees and we can't get something for nothing," he said at the time. "Talk is cheap. Roads are not."

A British man named Adam Greenfield comes to the microphone. "Austin is an amazing place," he says. "It is a beautiful city. We have everything here. And"—his voice lowers—"we have I-35. A dangerous, polluting, dead zone in the heart of what should be thriving, bustling neighborhoods. No city needs an interstate highway running through its downtown and we don't either." He pauses dramatically while two women standing behind him unfold a giant banner showing a rendering of I-35 as a tree-lined city street, clogged not with cars but with people and bikes and buses. "It's time to remove I-35 through Austin and replace it with a boulevard. This can be done!" he says. "It has been done. Rochester, New York, is doing it right now. We can let TxDOT keep making the same mistakes, or we can create a spectacular boulevard, one that is uniquely Austin." The driver of a semitruck lays on its horn overhead, drowning out the smattering of applause.

Adam grew up in Guernsey, an island slightly larger than Manhattan in the English Channel, just off the coast of France. "You can't drive longer than seven miles in any direction before you fall into the ocean," Adam says. The widest road on the island had only two lanes. After college, he met an American woman from San Francisco and fell in love. A few days before he was going to move to California, he got a call. His girlfriend had been killed in a car crash. He was devastated—and decided to move anyway.

As he settled into his new city, Adam set up a Facebook group for his neighborhood and decided to organize a few events to justify the group's existence. "And I realized, where do we actually go to get together? There wasn't a good answer to that question," he says. Bars were no good; people wanted to bring their kids. The closest park felt too far. A restaurant was loud and expensive. "I realized that the built environment profoundly impacts our entire lives, the relationships we have with our neighbors, how we feel when we move around," he says. He started organizing street fairs and block parties in neighborhoods across San Francisco. The first block party he organized, he chose a street almost at random and knocked on every door, asking the occupant if they wanted to have a block party. It felt as though people had just been waiting for someone to ask that question. In the months after the party, he watched "an explosion of community" on that one block. "It had a profound effect on their relationships with each other and their ability to organize," he says. He moved onto other blocks. Some parties flopped, of course. But Adam saw the potential. "We use streets for this one task most of the time, which is moving, and particularly moving cars, but it doesn't have to be that way. We could use streets for many things," he says.

He didn't have a car, so he biked around San Francisco, "completely oblivious to danger." He didn't plan his routes, didn't go out of his way to find a protected bike lane. He just went where he was going. And then, one August night in 2014, he showed up at a Critical Mass ride. As he rode around the streets of San Francisco surrounded by hundreds of other cyclists, he felt safe, which made him realize how *unsafe* he'd felt on every other ride he'd been on before. "That was the first time that transportation became politicized for me," he says. He realized, "When lots of people are together, they can make things different."

By 2016, Adam had married an environmental scientist. When she accepted a job at St. Edward's University, a liberal arts college in Austin, Adam had never even visited the city. "I came here basically sight unseen," he says. He showed up at the office of Bike Austin and was soon helping organize campaigns to get protected bike lanes installed on city streets. Through his advocacy, Adam met an urban planner named Heyden Black Walker who invited him to monthly meetings at Black + Vernooy, an urbanist architecture firm founded by her father, Sinclair Black.

Sinclair had graduated from the University of Texas with a degree in architecture in 1962, the year I-35 opened to traffic. In 1969, he moved to Berkeley to earn his master's degree in architecture. At the time, the Bay Area was the epicenter of the first wave of freeway revolts. "That's when I first ran across the planners who were opposed to the Embarcadero," he says. "I came back to Austin and watched the damn elevated being built." It seemed absurd that as San Franciscans were agitating to remove an elevated highway, TxDOT was building a new one—and right through the middle of Austin. Over the following decades, Sinclair helped bring new urbanism to the suburban-feeling Texas city. In 2001, Black + Vernooy wrote Austin's downtown streets master plan, which articulated a vision of "streets for people" and outlined a strategy to rebuild dozens of city streets downtown to "calm traffic" and "change the space and scale of the street to create a sense of place for the individual." Austin would ultimately apply that plan to a grand total of three city blocks. "If I ever write a book," Sinclair said in 2017, "it's going to be called, *Austin, Texas: Lost Opportunity National Park*."

In the mid-1990s, TxDOT showed up in Austin with what would become the first of many plans to fix I-35. The renderings the agency published showed a highway with towering off-ramps "rising up to 20 feet in the air before swooping directly onto downtown streets," wrote Chuck Lindell, a reporter for the *Austin American-Statesman,* in 1996. Sinclair was offended. He sketched a plan to depress the highway belowground and cover most of it with a boulevard and parks, stitched into the street grid. Sinclair's plan would create "enormous development potential on the periphery of the sunken freeway," Lindell wrote—"land that is now 'poisoned' by the noise and inaccessibility of the elevated interstate."

By the late 1990s, TxDOT had abandoned its soaring off-ramps, and Sinclair went back to teaching at the University of Texas. But in 2011, TxDOT returned to I-35 with new vigor. Sinclair's daughter, Heyden, was an urban planner, although she had spent the previous decade teaching prekindergarten at a local school. Sinclair recruited her to come work at Black + Vernooy, and together they refined his original vision and called their campaign Reconnect Austin. They were inspired by Klyde Warren Park, a five-acre deck park that had just opened over the sunken Woodall Rodgers Freeway on the north edge of downtown Dallas. Suspended over eight lanes of highway, the park held more than three hundred trees, two buildings, a water fountain, and the capacity for ten thousand people. The park was a miracle, a conjuring of public space from a polluted highway, a suture over a scar. Sinclair and Heyden wanted TxDOT to build Klyde Warren Park in Austin—but bigger, spanning four miles instead of two blocks. Reconnect Austin called on TxDOT to depress the highway belowground and cover it with a tree-lined boulevard, reminiscent of East Avenue. In the process, the city could reclaim thirty acres of land currently occupied by frontage roads and highway barriers. It was an enormous amount of valuable land—taxable land. The city could sell the land to developers and collect the hundreds of millions of dollars in tax revenue from market-rate development. Or it could partner with nonprofits to create affordable housing. In 2017, the Austin Strategic Housing Blueprint had identified the need to create at least sixty thousand units affordable to low-income households. Mostly, what had stymied the city from building all that housing was finding a place to put it.

For a decade, Sinclair and Heyden beat the drum of Reconnect Austin. Sinclair was by then a famous architect, and his status got him meetings with TxDOT engineers, who listened to him politely and then went about their business. He assembled a group of urban planners and architects, and they met regularly at the Black + Vernooy office. One evening in 2019, at one of those gatherings, Adam overheard someone say, "Why don't we just get rid of that stupid highway?"

"It hit me like a bolt of lightning," he says. Why wasn't there a campaign to tear down I-35? He used to lead bike tours along the Embarcadero in San Francisco. He knew how people had opposed that teardown: Merchants in Chinatown insisted it would kill their businesses; nearby residents feared traffic would flood their streets. "They were all dead wrong," Adam says. He was familiar with Patrick Kennedy's campaign to tear down I-345 in Dallas; he'd been impressed by the studies showing the financial benefit of highway removal. And Reconnect Austin was itself a compromise, an acknowledgment that TxDOT would never remove an interstate from the fastest-growing state's fastest-growing city. Adam talked to Heyden and Jay Blazek Crossley, and they agreed that there should be a campaign for full removal, shifting the Overton window on what was possible in the corridor.

In the fall of 2020, Adam biked from his house in East Austin to the end of Ninth Street, walked around a guardrail, and sat down in the damp grass to consider the highway at the bottom of the hill. The traffic thrummed below, a low roar of rushing cars and rattling 18-wheelers. Just beyond I-35, Austin's downtown glistened in a spread of glass and steel. On the other side of the highway, the Austin Police Department's headquarters towered over Eighth Street, the backdrop to the Black Lives Matter protests that had erupted in June, when thousands of people climbed onto I-35 and stopped traffic on the interstate that would eventually pass through Minneapolis, blocks from where the Houstonian George Floyd was killed by police. He felt the full force of the highway's effects on Austin. It was crushing. But he could also imagine it gone. "I was able to look at I-35 and see it as a decision that was made and not this thing that's always been there and is inevitable," Adam says.

He went home and started a campaign called Rethink35, which called on TxDOT to remove I-35 altogether. At first, the campaign focused on the boulevard that would replace the interstate, a street with space for cars but also protected bike paths and dedicated bus lanes and wide sidewalks, lined with apartments and offices and restaurants. But people couldn't wrap their heads around the idea that I-35 would just be *gone.* So he started talking about State Highway 130, which swung around Austin to the east. Nonlocal traffic should go there, bypassing Austin. "If you make the jump from there—if you say, well, once nonlocal traffic is going around town, then why do you need an interstate going through Austin?" People agreed: you didn't.

Removing the highway required investing in other ways to get people where they were going. Because state law required that roughly 97 percent of TxDOT's $16 billion annual budget be spent on roadways, transit in Texas cities was mostly funded through sales tax revenue. But in the fall of 2020, Austin voters had approved an increase in city property taxes to fund a $7.1 billion investment in public transportation called Project Connect, which would include two light-rail lines and a dozen new bus routes. Since 2000, Austin voters had twice rejected transit plans, and in the weeks leading up to the 2020 election, it seemed as if the vote could go either way. Property taxes were already high and getting higher, and residents bristled at the idea of paying more for housing in order to fund transit. But Project Connect had passed, and by a wide margin. Federal funding would cover roughly 45 percent of the construction cost, while the rest would be funded through a permanent property tax increase.

If expanding the interstate allowed Austin to grow bigger—sprawling deeper into the suburbs—the goal of Project Connect was to make Austin feel smaller. "My hope is that once we've built out our system, people don't really think that much about it. They just use it. And as a result, our city feels smaller because places are more connected and we have more places, more centers of activity," says Peter Mullan, the chief of architecture and urban design at Austin Transit Partnership, the nonprofit that was created to build and manage the new rail network. "In cities—even in a place like Austin, in a place that's changing as much as Austin is changing—people have a really hard time imagining things being different from how they are right now. Even though they're so different than they were five years ago. It's very hard to imagine a future other than the one you're in right now. When the reality is that the only constant in cities is that they change. You're either changing for the better or for the worse."

In August 2021, just before the public meeting at Huston-Tillotson University, TxDOT published a technical report that included an evaluation of three "community concepts," including Rethink35 and Reconnect Austin. According to TxDOT, neither alternative was feasible. "Although the physical rebuilding of I-35 into a boulevard with wide sidewalks, accommodations for transit, bicycles and pedestrians can be done," the report concluded, "the traffic impacts to the surrounding streets and delays it would cause to through traffic make it unlikely that such a concept would meet the transportation needs of an interstate highway." It was difficult to evaluate such a dramatic reshaping of the I-35 corridor, because "the ripple effects would extend far beyond vehicle and person travel." In its traffic models, TxDOT hadn't considered how Project Connect might reduce demand on the highway—how the new rail and bus lines might siphon away drivers, easing congestion without adding lanes.

"The final plan needs to do more than just pay lip service to transit," Natasha Harper-Madison had said while standing under the highway on that hot September morning. "It needs to complement rather than compete with our Project Connect investments."

One of the last speakers to come to the lectern is Jaime Cano, a soft-spoken man in his early forties. He introduces himself as the assistant director of Escuelita del Alma, a Spanish-immersion day care and preschool located just north of Stars Cafe on the I-35 frontage road. "If TxDOT proceeds with the proposed highway expansion, Escuelita del Alma child-care center will be forced to close its doors after twenty-one years of caring for and educating young children," he says. He is accustomed to speaking to three-year-olds; there is no anger in his voice as he leans into the microphone. He is professional, unemotional. "As a result, two hundred families will lose their trusted child-care center, which would add to the already severe shortage of quality child-care centers in the city." The school would be required to find a new space, one that meets strict state requirements for child-care centers. "In this current real estate market, it will not be possible to find a facility that is affordable or even reasonably priced," Jaime says. Likely, after two decades in business, the school will cease to exist and those two hundred families will have to find child care elsewhere.

In 2000, when Dina Flores opened Escuelita del Alma on Congress Avenue and Second Street in a one-story brick building next to a beloved Tex-Mex restaurant called Las Manitas, Austin was a very different place. The tallest building downtown was an office building called One American Center that rose four hundred feet over Congress and Sixth Street, an otherwise desolate intersection. Few people lived downtown, which emptied out after 5:00 P.M.

Dina grew up in Mexico, just across the border from Del Rio. She moved to San Antonio with her family when she was fifteen, eventually earning a master's degree in early childhood education and teaching kindergarten in San Antonio public schools, where most of her students were native Spanish speakers. Dina moved to Austin in her early thirties with her partner, Cynthia Pérez. When Pérez opened Las Manitas with her sister, Lidia, Dina worked as the restaurant's bookkeeper. Many of Las Manitas' employees needed child care, but there was nothing downtown. As a kindergarten teacher, Dina had seen the difference between children who were sent straight to school and those who had some kind of formal education beforehand. "Those children who had gone to child care were further ahead than the other children," she says. "So I just felt like, okay, it's time for me to start working with children at an even younger age than kindergarten." In 2000, Dina opened Escuelita del Alma next door to the restaurant and started caring for young children, infants through prekindergarten. She wanted to do dual-language education, but she had a hard time finding bilingual teachers. "At some point I decided, they're going to learn English regardless," she says. "This is the perfect time for children to learn another language." Escuelita became the city's first Spanish-immersion preschool and the only full-time child-care provider downtown. Soon, a hundred families were enrolled at the school with another hundred on the waiting list. Las Manitas provided daily lunches, and restaurant staff got discounted tuition for their kids. In October, the older children went trick-or-treating at downtown businesses, stopping by city hall and the Texas State Capitol, just up Congress Avenue.

One day in 2006, a lawyer representing a real estate developer stopped by Las Manitas during a busy lunch rush. He was there to give Cynthia and Lidia Pérez official notice that their building had been sold and would soon be demolished and replaced by a towering Marriott hotel. Cynthia called over the bustling restaurant, silencing lunch service. "Ladies and gentlemen, this is the lawyer who represents the white guys who plan to bulldoze Las Manitas!" (The lawyer responded that he also wanted a taco.) The community rallied around the little restaurant and school. Austin's city council wrote a letter to Marriott's chairman and CEO stressing the cultural and historic value of Las Manitas and Escuelita del Alma. "It is a source of pride that these institutions have grown up locally and thrived on Congress Avenue, one of the most high-profile, historic streets in Texas. These businesses are . . . what makes Austin a unique and special place. Losing them is simply a losing proposition for every-one involved."

But the land had already been sold. The developer didn't need city approval to build a high-rise hotel. "There was nothing that could be done," Dina says. She spent a year and a half searching for a new lo-cation for the school. Parents hosted fundraisers. Maybe they could buy a building? The banks all said no. "The two businesses that nor-mally fold within the first three years are restaurants and child-care centers," Dina says. So she kept searching.

One morning, six months before she had to vacate the building on Congress Avenue, a parent called her. He'd just seen a FOR LEASE sign on a building on the I-35 frontage road near Thirty-Second Street in Cherrywood, the neighborhood where he lived. Maybe she should check it out? The next day, Dina was standing outside the low-slung limestone-brick building, looking up at the elevated high-way that clattered above it. Inside, the building was a maze, bigger than it appeared on the street, its hallways branching into a dozen rooms. They would have to retrofit the building to make it work as a child-care center—install tiny handwashing sinks and toilets in every classroom, build two playgrounds on the patio—but otherwise it was perfect. There was plenty of space to expand and accommodate more students. Dina was relieved: "It's going to work."

There'd already been a lot of interest in the building, the leasing agent told her. It was owned by the movie director Richard Linklater, who had become famous for his work on the iconic Austin movies *Slacker* and *Dazed and Confused*. Linklater had bought the building in 1993, the year *Dazed and Confused* was released; he needed produc-tion space and the sprawling building next to I-35 was affordable. "You got used to the freeway noise pretty quickly," Linklater told me. "Every now and then you'd hear the squealing of brakes fol-lowed by crunching metal due to the extremely short entrance ramps they had at that time."

The leasing agent told Dina that Linklater was out of the country and that she probably wouldn't hear back on her application for a few weeks. In late February, while she was still waiting to hear about the building, Dina got a call from a woman looking for child care for her nephew. The woman's sister was moving to Austin from California, where she worked as a teacher, and needed someone to care for her son when the new school year started in August. At that point, Dina had no idea if she'd still have a school in August. "I told her, I can guarantee you that we will have a spot for him in August. But I don't know if you're aware that we're the child-care center that is going to be displaced," she said, referencing the almost weekly news stories about the future of Las Manitas and Escuelita. She'd just put in a bid for a new building, Dina told the woman, but she didn't yet know if they'd gotten the lease. The woman asked where the new building was located. "It's right on the northbound access road, at 35 and East Thirty-Second Street," Dina said. "And she goes, 'Oh, my husband owns that building.' "

"Your husband is Richard Linklater?" Dina asked. "And she says, 'Yeah.' And I said, 'Well, can you please tell him to give us the lease so we can have a place to move to?' " He was out of town, the woman said, but she'd talk to him when he got back.

Dina got the lease. Richard Linklater's nephew got a school.

Moving was a nightmare. There was the remodel, retrofitting the building so that it conformed with state requirements for a child-care center. When it was finally ready for children, Dina was determined to not close the school and leave parents scrambling for care. For weeks leading up to the move, teachers boxed up their classrooms, carefully labeling each box of toys and markers and books. One Friday in December 2008, parents picked up their children on Congress Avenue for the last time. Volunteers poured into the school, loading desks and chairs and shelves into pickup trucks and vans and drove north on I-35, arriving at the new building to find Dina directing the flow of boxes and furniture—this here, that over there.

After the turmoil of the move, Escuelita gradually settled into its new home. Families dropped their kids off in the morning and picked them up in the evening. Little kids became bigger kids. Families gathered at the school for graduation in the summer and Halloween in the fall. Instead of downtown businesspeople and bureaucrats, parents worked at the hospital across I-35 or at the University of Texas, a few blocks west. In the larger space, Dina was able to more than double the school's capacity to 200 students and hire more teachers. After a few years, they leased another building across the street for their infant care. They closed briefly when COVID-19 swept through the state. When they opened again, kids showed up at school in tiny masks.

By early 2021, Dina started to think about retiring. She was sixty-six and she was tired. Her daughter, Alma—the school's namesake—had returned to Austin and worked as an assistant director, along with Jaime Cano. Maybe they could take the school over and continue the work that Dina had dedicated her life to.

TxDOT had been talking about expanding I-35 since Dina had moved into Richard Linklater's building. "I would always read that at some point the highway was going to be expanded," she says. So she didn't worry much about it. But then, in the summer of 2021, a former parent called Dina. Jim Walker's son had attended Escuelita when it was located on Congress Avenue and they'd stayed in touch. Jim had moved to Cherrywood in 1995 and dutifully showed up at public meetings every time TxDOT announced a new plan to reconstruct I-35. This time was different. The project was funded, Jim told her. It had momentum; it had schematics. Those schematics showed the highway running straight through the Escuelita building. "He said, 'This time it is going to happen,'" Dina recalls. She needed to find another location for her school.

Not again, she thought.

8

PAUSE

Houston
I-45

Throughout 2020, consultants hired by TxDOT continued to contact Modesti, offering to buy her house. Appraisers showed up on her doorstep, asking to see inside. Surveyors started marking lots on her street, wood sticks fluttering with bright pink flagging tape appearing in front yards. Sometimes it felt as if the offers were more like threats: She was going to have to give up her house eventually, they said. Why not do it now, when they would help her move?

One day, she let an appraiser inside to see her house. They walked upstairs from the entryway to the second floor, the heart of the house, with an open-concept kitchen and living room. The house was clean and bright, thoughtfully decorated, with hanging globe lights over a marble kitchen island and two gray leather couches arranged in an L shape in the living room. Spanning the room's north wall were three square picture windows looking out over Interstate 10. The highway seemed to fill up the view, cars streaming along its cement expanse. Modesti and the TxDOT appraiser climbed a flight of hardwood stairs to the third floor and peeked into the master bedroom and a guest room with two bunk beds for her godsons. On the fourth floor, they walked out onto her narrow balcony, where she watched fireworks cascade over downtown on the Fourth of July.

She showed him the den, where she watched movies with her family on a flat-screen TV arranged under a navy ceiling studded with tiny LED stars. A window in the northwest corner of the room framed a view of the rooftops below, lined along the highway frontage road. A fully stocked bar glinted in the sunshine, lined with dozens of unopened glass bottles. When Modesti lived overseas, every time she visited a new country, she would buy a bottle of its national liquor and mail it back to Houston. She didn't drink much, so it was more about creating a collection, documenting her travels. The northern wall of the navy room was decorated with medals, commendations, and other memorabilia from the military missions she served on.

"You really put a lot of thought into this house," the TxDOT appraiser told her. Of course she had. She'd built it from the ground up, considering every detail. "If I am going to be here, I am going to make it a part of me, and go from there," she says. TxDOT would pay her for her house, but not nearly enough; the first offer the agency gave her was $50,000 less than her home's appraised value when she moved in, she says. It wasn't enough to start over, to build another house on a similar piece of land.

Since Modesti bought her property in 2015, gentrification had swept through the southernmost stretch of the Fifth Ward, the neighborhood sandwiched between Interstate 10 and Buffalo Bayou, known as the Bottom. When Modesti bought her land, her street was full of empty lots, scraped of the densely packed single-story pier-and-beam houses that had once made up a neighborhood. But developers had found this vacant land, and three- and four-story townhomes started sprouting throughout the Bottom. A company called Midway bought a giant swath of land extending more than a mile along the bayou and announced plans for a multibillion-dollar, 150-acre, mixed-use development called East River. "As soon as my house was done, everything behind me came," Modesti says. She remembers thinking, "Whoa, I'm going to have a lot of new neighbors." She'd bought her land for $50,000. Now lots were selling for $200,000.

The longer Modesti lived in her home, the more she grew to love her neighborhood. Her neighbors invited her to cookouts and parties; whenever she ventured out, someone or another would be outside, hollering across the fence or street: "Where you going, neighbor?" Modesti lived in one of the only new homes on her block. Many of her neighbors had been in their houses, built in the 1930s and 1940s, for generations. Modesti says she thinks TxDOT chose a neighborhood that "they thought was in distress. I think it's easier for them to say, okay, well, their homes are old and rag-gedy. We should give them some money. They can buy a new home."

A block down from Modesti, Sean Jefferson lived with his family in one of those "raggedy" houses, built in 1940. It wasn't fancy, but it was a decent place to live, and it was affordable; he'd paid $64,000 for the 850-square-foot home a decade earlier. "TxDOT hasn't did anything but come and do a survey and send a letter in the mail, say-ing that we're going to be taking this land for a freeway project," he said in a video about the expansion. "They had someone come to a meeting, but they are just explaining the process of what's going to happen. Not trying to get any kind of community input. . . . When they come to do the groundbreaking, home is not going to be home."

Slowly, people started to leave, accepting the offers they'd been given. A few doors down from Modesti a family of eight lived in a single-story home with their German shepherd. "One day I saw them pack up in a U-Haul and just leave." She'd liked having the Ger-man shepherd around, hearing it bark whenever someone new passed by. Now the house was vacant. "I feel that they wouldn't go to homes over half a million dollars and demand them to leave," Modesti says. "They see these homes as run-down; they can't feel bad because they aren't new. My thing is, they are historic. You don't knock them down, you revise." It felt discriminatory. TxDOT wasn't going to wealthy white neighborhoods and kicking those people out of their homes, she says. Indeed, the only neighborhood where

TxDOT had considered right-of-way impacts when advancing early alternatives for the project was in the Heights, an affluent area north of downtown.

As Stop TxDOT I-45 mobilized a grassroots opposition to the expansion, people started showing up at city hall, demanding that their elected officials do something. In response to public outcry, Houston's mayor, Sylvester Turner, hired an outside consulting firm to do a new round of public engagement—this time, for the City of Houston instead of TxDOT. In the spring of 2020, Turner presented TxDOT with the city's vision for the project, stressing that I-45 should be rebuilt within its current footprint, with dedicated bus lanes, more crossings, and space for cyclists and pedestrians. TxDOT ignored it. In December 2020, Turner again sent a letter to TxDOT, stressing that the highway should be rebuilt but not widened and threatened to pull the city's support if the project's "shortcomings" were not addressed. The Houston-Galveston Area Council, the metropolitan planning organization that had voted to fund the expansion almost two years earlier, had also cooled on the project. That winter, members spent months wrangling a memorandum of understanding, or MOU, that would hold TxDOT accountable for mitigating the adverse impacts of the expansion. TxDOT refused to sign it and instead insinuated that if Houston didn't like the project, the agency would go ahead and take its $7 billion elsewhere. "If there should be a desire to change the nature of the project, we should follow the federally defined processes to make those changes and not attempt to do so through Resolutions or MOUs," said TxDOT's executive director, James Bass, in a written statement to the council.

By early 2021, TxDOT was on the verge of finalizing the project by issuing a record of decision, the final step in the federal environmental review process before construction could begin. Under the 1970 National Environmental Policy Act, or NEPA, any state agency receiving federal funding for a project had to document how that project would affect the human and natural environment. Before 2014, when TxDOT wanted to expand a highway, it wrote up an extensive environmental impact statement—which by law included several rounds of public comment—and then submitted that document to the Federal Highway Administration and waited for approval. But the federal agency was understaffed, with fewer than a dozen people reviewing Texas's highway projects. And TxDOT had a lot of projects. So in 2014, the Federal Highway Administration gave Texas the authority to enforce its own NEPA compliance—essentially, to self-certify that it had followed the law. California already had the authority to self-certify its environmental review; six other states would eventually join the program, called NEPA Assignment. Although the arrangement allowed the state to move more quickly on projects, reducing cost and unnecessary delays, it also meant that TxDOT operated with essentially no federal oversight.

On February 4, TxDOT issued its record of decision for the North Houston Highway Improvement Project. "The future of transportation is changing and the infrastructure in the nation's fourth largest city needs to change with it," TxDOT announced. This project would help Houston prepare for the future "by improving resiliency to weather events and providing safer, more efficient travel." The highway expansion was moving forward, and Houstonians needed to either get on board or get out of the way.

Meanwhile, a thirty-three-year-old named Christian Menefee was settling in as the new Harris County Attorney, the first Black person to hold the office, as well as the youngest. His dad, a veteran, had grown up in the Fifth Ward, and Menefee had campaigned on holding corporate polluters accountable. He'd been on the job a little over a month when TxDOT approved the highway expansion, and his staff started talking about filing a lawsuit against the state agency. They thought there were issues with TxDOT's environmental review—that the agency had violated NEPA by adopting a design that "ignored serious harms, disregarded the concerns of the communities impacted by the Project, and brushed off the numerous comments they received as part of their flawed EIS process," as they alleged in the case they ultimately filed on March 11.

That same day, Menefee, Harris County Judge Lina Hidalgo, and the county commissioner Rodney Ellis convened a press conference. "For too long, transportation policy in our region has been stuck in the fifties," Hidalgo told a roomful of reporters. "For a generation, we've gone on just building more lanes, putting down more concrete, thinking that somehow magically that's going to reduce traffic and that's going to make us more competitive. All the while, what we've done is created more flooding problems, exacerbated instead of solving traffic, and built bigger and wider highways." The city and the county had tried to work with TxDOT, she said. "We have convened meetings. We've proposed memorandum of understanding. But time after time TxDOT has done nothing more than give us and our community lip service. They've made promises they can't keep, bulldozing right through us with this process."

No one expected the lawsuit would ever be considered in court. NEPA was a procedural law, mandating the process that a public agency must follow rather than requiring any particular outcome. A district court judge—in Texas of all places—was unlikely to tell TxDOT to stop its expansion. The lawyer they hired to help file the lawsuit told Menefee, "Anything that you get for this project is going to be as a result of negotiations," Menefee later recalled. He knew he had little leverage. "No court is going to come in and do anything other than say, 'Hey, TxDOT, cross this t, dot this i.'" Menefee hoped to force TxDOT to the table—to obtain some kind of meaningful concession that made the project better.

But then Menefee got good news.

In late February, as Harris County was preparing its lawsuit, Modesti filed a complaint with the Federal Highway Administration. In the complaint, she alleged that the expansion project violated Title VI of the 1964 Civil Rights Act, which prohibits discrimination "on the basis of race, color, or national origin" in any project that receives federal funding. Unbeknownst to Modesti, Bakeyah Nelson, the executive director of the nonprofit Air Alliance Houston, had already submitted her own complaint, co-signed by sixteen community leaders and grassroots organizers, including Molly Cook. Construction of the original I-45 project had already harmed historically Black and Hispanic neighborhoods, Nelson wrote to Bass, TxDOT's executive director. Widening the highway would only "repeat and perpetuate the wrongs of previous siting and expansion decisions and the decades of significant adverse impacts on these neighborhoods." Approving the project, she alleged, would violate the Civil Rights Act.

On March 8, the Federal Highway Administration sent a letter to Bass. "We're writing with regards to three recent letters received by the Texas Department of Transportation and forwarded to the Federal Highway Administration," it began. Those letters raised concerns that the project could be in violation of Title VI of the Civil Rights Act. "To allow FHWA time to evaluate the serious Title VI concerns raised in the letters . . . we request that TxDOT pause before initiating any further [activity] for the project," the federal agency wrote. If FHWA found that discrimination had occurred—that the highway expansion "creates potential disparate, adverse impacts to the predominantly African American and Hispanic communities within the project area"—it could stop the project entirely.

The letter sent a shock wave across Houston: The federal government had intervened to pause a massive highway expansion. It was unprecedented—"a really huge deal," said the former head counsel at FHWA. Menefee was thrilled; suddenly Harris County had leverage. Advocates were overjoyed. That Friday, Molly threw a party for activists and community members in her backyard.

One day in April, a month after the project had been paused, a woman came by when Modesti was outside mowing her lawn. The woman said that she had been hired by TxDOT to help Modesti and other homeowners transition smoothly out of their homes. "You know the expansion is coming. You need to move," she told Modesti. But Modesti wasn't moving. Didn't the woman know that the project had been paused? A few weeks later, Modesti checked her mail and found a postcard with an image of Che Guevara on the front. The message on the back was written in looping cursive with a black ballpoint pen. "Thank you so much for taking time out of your busy day (mowing your lawn) to talk with us. I want to acknowledge your service to America and thank you. I also want to tell you that I understand you championing the cause and filing a Civil Rights complaint with FHWA. I respect you for that. Please understand our role came about because of Civil Rights and Environmental Justice." Another one, a week after that: "Please let me help you. Please call me. I so want to be of service!" It was signed "Del Richardson, your Chief Advocate." Del Richardson & Associates, or DRA, was the name of a company hired by TxDOT to provide relocation services in the Fifth Ward.

The postcards creeped Modesti out. She sent photos to Susan and Molly and then contacted FHWA—she thought they had paused the highway project? They had. In June, FHWA sent another letter to TxDOT to clarify what, precisely, they had meant by "pause." The federal government had meant *stop:* "FHWA believes that no further actions be taken on this project that might impact our Title VI investigation and any proposed remedies should the agency find that a violation has occurred." Tucked into the final paragraph was another update: In addition to its civil rights investigation, FHWA would be reviewing TxDOT's compliance with the NEPA Assignment program. If the state was found to be out of compliance, TxDOT could no longer approve its own environmental reviews—for projects across the state.

Until the project was paused, Modesti thought she was the only one who had contacted the federal government. "When I filed my complaint, I asked the gentleman, had there been any other complaints? And he said, no, I was the only one," she says. When she saw the pause letter, Modesti was relieved. But still, she couldn't relax. "I've been just praying and hoping that they do try to listen to us. From what I'm hearing, everyone is like, TxDOT is going to do what they want to do. This might stop them, but they'll come back."

PART II

9

TRANSIT

Washington, D.C.

On a cold, overcast afternoon in late March 2021, President Joe Biden walks into the Carpenters Pittsburgh Training Center, a trade union and school on the outskirts of the city. Standing in front of a forest of American flags, in an echoing warehouse lined by plywood panels and two-by-fours, Biden announces his $2.3 trillion infrastructure plan—"a once-in-a-generation investment in America, unlike anything we've seen or done since we built the Interstate Highway System and the Space Race decades ago." Through a historic investment in transportation infrastructure, Biden intended to modernize America's "crumbling" roads and bridges, invest $85 billion in public transit, repair and expand passenger and freight rail lines, and incentivize electric vehicle production. But it wasn't just about infrastructure; this was about jobs. Good jobs, millions of them. Infrastructure investments had long been bipartisan, Biden says, often led by Republicans. "I don't think you'll find a Republican today in the House or Senate—maybe I'm wrong, gentlemen," he says, glancing over to his right at some unseen Republican delegation in this union hall—"who doesn't think we have to improve our infrastructure."

The question, of course, was, what kind of infrastructure?

Two weeks later, Senator Sherrod Brown convenes a virtual hearing for the Senate Committee on Banking, Housing, and Urban Affairs. The FAST Act, the surface transportation bill that had passed in 2015 under the Obama administration funding highways, transit, and rail, was set to expire in the fall. The intent of this hearing is to discuss funding for public transportation in the upcoming surface transportation reauthorization. Despite the president's ambitions for a bipartisan infrastructure bill, the hearing is decidedly partisan. "All Americans should have high-quality, frequent transit service that saves them time and money," begins Brown, a Democrat from Cleveland, Ohio. "When you have better, faster transit service, more people use it . . . a vibrant, fully functioning transit system is essential to attract investment and create communities where people want to live and to work."

The ranking Republican on the committee is Pat Toomey, a second-term senator from rural Pennsylvania. Biden's proposed $85 billion transit investment was, Toomey says, wasteful and excessive. Add that to the money the federal government had already spent supporting transit agencies through its COVID relief packages, and the government "could buy every transit commuter in America a $30,000 car. . . . We need to ask a tough question, which is does every mass transit system in America make sense at its current scale?" he says. For David Ditch, a research associate at the Heritage Foundation, a conservative think tank, the answer is a resounding no. Using the Highway Trust Fund to pay for transit is fundamentally unfair, he insists. "Users of highways and airports directly cover most of the cost of highway and airport infrastructure. In contrast, transit users cover less than a third of the cost of transit systems, which receive a variety of subsidies," he says. "As such, the mass transit account represents a longstanding wealth transfer from people who drive to people who do not drive."

This is not true. In 2008, when the U.S. Department of Transportation announced that the Highway Trust Fund had run out of money, Congress responded by bailing out the trust fund with $8 billion from the U.S. Treasury. Since then, general taxpayers—drivers and nondrivers alike—have paid more than $140 billion to subsidize the Highway Trust Fund. This fact does not deter Ditch from his message: Highways are paid for by users, and this represents good free-market capitalism. Subsidized transit is socialism.

Toomey addresses his other witness, Baruch Feigenbaum, a managing director for transportation policy at the Reason Foundation, a libertarian think tank. "If systems do choose to expand into, say, an affluent suburb, is it reasonable to ask affluent people in the suburbs to pay the cost of their ridership?" Toomey asks. It is a leading question, like all questions at this hearing. Yes, of course, Feigenbaum says. "If we want to expand service to transit-choice riders, commuter rail, for example, it is absolutely fair to ask those folks to pay for it . . . I don't think federal taxpayers should be subsidizing the costs of these systems."

Bob Menendez, a Democratic senator from New Jersey, jumps in with a rare unscripted remark. "Well, we subsidize highways. We subsidize bridges. We subsidize all those things in the investments we make. I'm not quite sure what is the difference," he says, with a quick shake of his head before he regains control of the script. One of the witnesses in attendance today is Beth Osborne, the director of the nonprofit Transportation for America and the former assistant secretary for transportation policy at the U.S. Department of Transportation. Menendez directs his first question at her. "In recent history, when working on surface transportation reauthorizations, we've adhered to an eighty-twenty rule, of 80 percent spending on highway, 20 percent goes to transit," he says. "Ms. Osborne, your organization has advocated for a much more aggressive fifty-fifty split for highways and transit." Why, he asks, should the government make such a significant shift?

Beth appears on the screen before a blurred-out background, her gaze sharp behind bright blue oval glasses. "We have never made the kind of investment in transit at the national level that we have in highways," Beth says. "Transit is absolutely essential to a functioning economy." Later, she will double down: "If we really want to commit to transit, it's time to commit in the way we did [with] highways seventy years ago. . . . Americans travel elsewhere and ask why they have such good transit service there. Why can't we have it? And the answer is we haven't made the same level of commitment." Months earlier, more than thirty Democrats had called for such a commitment to transit, introducing a bill that declared "that public transit is a national priority which requires funding equal to the level of highway funding." The bill was Beth's idea. It didn't have a chance of passing—it didn't even get a hearing—but she wanted to raise the issue of the unquestioned hegemony of highway spending. "Transit is in high demand and low supply," she says now.

Until the 1970s, transit in America was considered a private, for-profit enterprise. Early transit systems were built by business owners or real estate developers; the profit was not in the journey but in the destination. The first transit system reportedly emerged in western France in the 1820s, when a retired army officer hired stagecoaches to bring people to his public bathhouses outside Nantes. When he noticed that people were getting off midway to the baths, he expanded the frequency and number of routes, eventually calling his coaches omnibuses. In the United States, as people poured into northern cities in the early twentieth century, horse-drawn streetcars—and then later electric streetcars—expanded the space available for settlement. "Mass transit was a centerpiece of urban life," writes the historian George Smerk in *The Federal Role in Urban Mass Transportation*. "It was almost the universal means of urban transportation in a time virtually devoid of private transportation."

The appeal of the automobile was obvious; it carried you wherever you wanted to go, whenever you wanted to get there. Transit ridership peaked in the early 1940s and began a steady decline as the automobile ascended. The prevailing view was that if transit companies happened to lose money, well, that was their problem—a signal from the market that their services were no longer required. And the government didn't interfere with the market by subsidizing private companies. "Because it had always been private, there was little concern over transit as an activity for the government to take action about other than to keep fares low and to play some regulatory role," Smerk writes.

But of course, the federal government *had* subsidized some private companies, and lavishly so—the companies that built highways and the ones that sold the cars that filled those highways. Transit companies couldn't compete with the enormous federal subsidy given to road builders. And the pattern of growth those highways enabled—dispersed, sprawling suburbs—was antithetical to bus and rail systems, which required a dense concentration of riders to be profitable. According to the American Public Transit Association, there were 17.2 billion passenger rides nationally in 1950. By 1960, that number had plummeted to 9.4 billon. "These passengers aren't lost," the transit association said in 1973. "Most are riding in from suburbia, one to a car, and sitting bumper-to-bumper on the most magnificent highways money can buy." Facing declining ridership, many transit companies went out of business. By 1973, 268 mass-transit companies "have gone out of existence," reported *Reader's Digest*. Another twenty faced bankruptcy.

In 1932, the federal government started collecting a cent per every gallon of gasoline sold. Until the Highway Trust Fund was created, revenue from the gas tax—which exceeded $30 billion between 1932 and 1957—was funneled into the government's general fund and used for a broad array of programs, much like the income or sales taxes. But after the Interstate Highway Act passed in 1956, revenue from the gas tax was diverted to the Highway Trust Fund and used exclusively to pay for the construction of the interstate highway system. It was a tidy, self-contained form of financing: Highway users—drivers—would pay taxes on fuel, as well as excise taxes on trucks and tires and other manufactured goods, and those taxes would be used to build the highways those users drove on. Fair and square.

The Highway Trust Fund was set to expire in 1972, once the interstate system was built out. But once the "highway-auto-petroleum complex," as *The Wall Street Journal* called the industry, got a taste of dedicated, unquestioned funding, it wasn't about to give it up.

In the early 1970s, facing the collapse of the nation's transit companies, advocates and lawmakers began to agitate for federal support for urban transit systems—to make transit a public enterprise. "Critics of the automobile's long supremacy in the nation's transportation system are out to break the Trust Fund," reported *The New York Times Magazine* in 1972. "They believe that this easy, automatic flow of money to the states from the Trust Fund for the sole purpose of highway construction has a profoundly distorting effect on the nation's transportation priorities and, indirectly, on the present and future development of cities and their suburbs."

That year, as the Federal-Aid Highway Act came up for reauthorization, some legislators began to wonder, shouldn't the federal government also be funding other forms of transportation? Didn't mass transit have a place in American cities? Two senators—Edmund Muskie, a Democrat from Maine, and John Sherman Cooper, a Republican from Kentucky—proposed that the Federal Highway Act be amended to give cities the option to spend the money they received from the Highway Trust Fund for bus or rail transportation. In 1972, for every dollar that was spent on mass transit—funded through a separate mass transportation program—the federal government spent $14 on highways. The Cooper-Muskie amendment proposed to balance the scales, allowing cities the option to spend federal funding on transit systems, money they would otherwise have to forgo if they elected not to build highways. "This bill would correct the most glaring flaw in our present policy—a closed circle of Federal funding which perpetuates highway construction as virtually the only solution to transportation problems that most cities and States can afford to choose," Muskie said. For example, after elected officials in San Francisco voted against the completion of the

Embarcadero Freeway along the waterfront, the city lost the money that had been allocated to the project through the Highway Trust Fund; instead, it had to raise property taxes to fund a rail system for the city, which it called Bay Area Rapid Transit.

John B. Anderson, a representative from Illinois and the chairman of the House Republican Conference, testified on the floor of the House in support of the amendment. "Through the rigid and un-yielding insistence that this Nation must spend $5 billion a year, come rain or come shine, whether needed or not, for new highways, they may unleash a wave of popular revulsion that truly will bring the great highway machine to a halt," he said. The idea that the Highway Trust Fund should be used to pay for highways and high-ways alone—the notion that the program was "self-financing"— betrayed a misunderstanding of how government worked. Should the alcohol tax be used only for substance abuse rehabilitation pro-grams? he asked. The cigarette tax funneled to the Cancer Institute? There were more than thirty product-specific taxes that netted $10 billion in revenue in 1971, none of which were earmarked for specific activities. Instead, they were funneled into the general fund and spent on the many activities of a national government, decided by the lawmakers the people had elected to represent them. Why should the gas tax be any different? Ever the pragmatist, Anderson insisted that drivers who had no intention of ever boarding a bus or train should fund transit, if for no other reason than that doing so would siphon *other* drivers off congested roadways. "We are con-fronted with a new set of facts and conditions as we enter the decade of the 1970s that were dimly if at all perceived at the time the inter-state highway program was launched in 1956," he said. "At that time energy supplies were abundant, urban air pollution had not yet reached really serious levels, the great suburban migration had just begun to take on its current dimensions, the traditional transit indus-try was still healthy if not robust, and few of us were far-sighted enough to see that merely expanding the mileage of modern urban freeways would be a treadmill like proposition due to the increased traffic that we now know such routes inevitably generate. Yet, if we could not know these things then, we are aware of them now."

Anderson's testimony would prove unpersuasive. The Cooper-Muskie amendment passed the Senate but was rejected by the House. Congress adjourned without a new highway act. When Congress convened the following year, "there was an element of desperation," Smerk writes. Many states were running out of federal highway money; construction would soon be disrupted. "Compromise was necessary to get the federal highway funds rolling again, even if it meant opening the Highway Trust Fund to mass transportation purposes."

Under duress, compromise was reached. The Federal-Aid Highway Act of 1973 included a diversion of $800 million in Highway Trust Fund money to transit programs—the first time transit had received any portion of the federal gas tax. "The Act is not only a highway act," said President Richard Nixon at the law's signing ceremony. "One of its most significant features is that it allows the Highway Trust Fund to be used for mass transit capital improvements . . . The law will enable [cities] to relieve congestion and pollution problems by developing more balanced transportation systems where that is appropriate rather than locking them into new highway expenditures, which can sometimes make such problems even worse." In that same law, highway spending surpassed $18 billion.

Almost a decade later, Ronald Reagan defeated Jimmy Carter and took office under record inflation and a crushing recession. (That year, John Anderson ran for president as an independent, garnering nearly six million votes.) In December 1980, inflation had hit 12.5 percent. The cost of highway construction more than doubled even as its funding source, the Highway Trust Fund, barely budged. In addition to sweeping Reagan into office, the 1980 election had ended the careers of several powerful rural Democrats, including the chairmen of both the Public Works and Transportation Committee and the

Ways and Means Committee. Those positions were claimed by two urban Democrats—both enthusiastic transit supporters. By that point, pro-transit legislators had been fighting for more funding from the Highway Trust Fund for more than a decade. But under the Reagan administration, they found themselves in a distinct position, with the surface transportation program up for reauthorization and a highway program facing an enormous deficit due to inflation—and the most pro-transit Democratic caucus in history.

The gas tax hadn't been raised since 1959, when Eisenhower increased it to four cents per gallon. In November 1981, months after Reagan signed one of the largest tax cuts in American history, senior White House staff started talking, quietly, about more than doubling the gas tax. Reagan was, of course, vociferously antitax, but this increase would be framed as a "federalism initiative," part of Reagan's goal to turn federal programs over to states. Provide an influx of cash as a bridge, and states could run their own transportation programs, managing the construction and maintenance of roads as they saw fit. "Right now there is a sizable Federal gas tax," Reagan told a reporter in 1981. "The Federal gas tax came into being for the Federal interstate highway system—and I'm wondering if that's ever going to be completed—and it was supposed to be a temporary thing."

Reagan didn't succeed in turning the federal transportation program back over to states, in part because states did not really want it back. But the promise of dismantling a federal program opened Reagan to the possibility of a tax increase. By 1982, there was bipartisan support in Congress for increasing the gas tax, which seemed to be the only option if the trust fund was to remain solvent. But the pro-transit urban Democrats who had recently taken office were skeptical; an increase in the gas tax would disproportionately affect low-income people, who spent a greater share of their income on transportation, specifically fuel. To garner their support, Secretary of Transportation Drew Lewis promised to create a mass transit account in the Highway Trust Fund, the first time there would be a dedicated carve-out from gas tax revenues. This mass transit account would receive 20 percent of the revenue from the proposed tax hike—a penny of the five-cent increase. With Congress on board, Lewis just had to convince Reagan. A brutal midterm election did much of that work for him. Republicans lost twenty-seven seats in the House of Representatives and Democrats promised to return to work and focus on job creation and economic recovery. "We cannot leave the field to the Democrats and with 10.4 percent unemployment the job issue will not go away," Reagan's advisers told him. "It does meet a genuine need (rebuilding our infrastructure), it is self-financing (although entailing a tax increase/user fee), and it does enjoy broad bipartisan support in the Congress." Finally, Reagan capitulated. The Saturday after Thanksgiving, during a radio address to the nation, he announced his proposal: a five-cent-per-gallon increase in the gas tax. "The cost to the average motorist will be small, but the benefit to our transportation system will be immense," he said.

It was a huge win for public transportation. But it would be the last big win for decades. "The dedicated revenues acted as a spending ceiling as much as a spending floor," says Jeff Davis, a researcher at the Eno Center for Transportation. With few exceptions, adjusted for inflation, federal spending on public transportation has remained essentially flat since 1982. The gas tax has since doubled to 18.4 cents a gallon. But the ratio agreed upon forty years ago—a political deal for a tax increase that has long since become obsolete—has remained. Highway programs receive 80 percent of federal transportation funding, while transit gets only 20 percent.

And we have gotten what we paid for.

By the summer of 2021, as the House and Senate considered two versions of an infrastructure package, Beth Osborne was moderately hopeful that the spell of the 80-20 split might finally be broken. She'd spent weeks fighting with legislators from both parties, and some had agreed to consider increasing funding for transit to 30 percent of Highway Trust Fund revenue. "That's an increase and very exciting," Beth says. "But I don't know that they'll stick there." Months later, when the $1.2 trillion infrastructure package finally passes both chambers of Congress, the status quo remains intact: 82 percent has been allocated to highway programs and only 18 percent to mass transit.

I meet Beth for lunch at a Chinese restaurant a few blocks from her house in Northwest D.C. in November, days before Biden signs the Infrastructure Investment and Jobs Act into law. She would be irate if it wasn't all so predictable. "The infrastructure bill is supercharged status quo, combined with a promise to use the programs that cause the problem to fix the problem," Beth says. "So much of what I think we've gotten wrong in transportation, ever since we started trying to reform it, is we always come up with a little sliver of money to fix the things that we're continuing to break with significantly more money."

Beth grew up in New Orleans in a family of five that owned four cars. A few years after she moved to Baton Rouge for college at Louisiana State University, she needed a job but didn't have a car to get to one, and she couldn't afford to buy a car without a job. Transit didn't take her where she needed to go and she wasn't about to put her life at risk riding a bike on dark streets with narrow shoulders. "My mom and dad bought me a car. Because I come from a family wealthy enough to do that," she says. "Every time I hear someone say, 'Americans love their cars,' I realize what they are really saying is, 'Americans love not starving to death.' I do too. If you don't have a car, you won't work. You won't get to the doctor; your kids won't get to school."

After graduating from LSU with a law degree, Beth moved to D.C. to work in education policy, inspired by her grandmother, an advocate who fought for desegregation. But with no experience on Capitol Hill, she couldn't find a job. To save money, she got rid of her car and eventually got hired as a legislative assistant in the office of a congressman from Pennsylvania, managing the office's environment and natural resources policy. "The area that kept coming up as the biggest problem for the environment and natural resources is transportation. It's bad for the air, it's bad for the water, it's bad for habitats, it's bad for open space," she says. When Beth arrived on Capitol Hill, Congress was wrangling the Transportation Equity Act for the 21st Century, the 1998 reauthorization of the surface transportation program. "All we were talking about in Congress at the time was, how much more damage should we accept from transportation?"

After years as a legislative staffer, Beth was appointed by President Obama to work in the U.S. Department of Transportation, eventually becoming the acting assistant secretary for transportation policy. "Being in the halls of the U.S. Department of Transportation, you are pulled in forty different directions. It's very hard to keep your focus and do affirmative things and not just react to stuff all the time," she says. Working at the DOT was eye opening. She'd spent years on Capitol Hill working to pass legislation to fund programs focused on equity and sustainability. Only once she got to DOT did she realize she'd been working on the margins. The Highway Trust Fund was the center, sacrosanct and untouchable. Through formula funding, billions of dollars flowed directly to states to do with what they pleased. The 1998 surface transportation reauthorization had created a formula funding system that guaranteed states would get back just over ninety cents for every dollar they paid into the Highway Trust Fund through fuel taxes. Until and unless the federal government put conditions on that funding, a state like Texas could do whatever it wanted, so long as it didn't violate civil rights or environmental law. If Texas wanted to expand every highway in the state to twenty-six lanes, it very well could. "We have a federally funded, state-run program," Beth says. "The states are the emperors."

At the Department of Transportation, Beth ran a $1.5 billion program called Transportation Investment Generating Economic Recovery, or TIGER, passed as part of President Obama's economic stimulus bill. The program allowed local communities to apply for grants directly from the federal government instead of going through their state departments of transportation. In 2009, when she was reading through hundreds of applications from across the country, she came across a project submitted by the City of Rochester. The city in upstate New York was proposing removing a portion of its Inner Loop, a sunken highway that circled downtown "like a noose, strangling the downtown area from adjacent vibrant densely populated neighborhoods," the city wrote in its initial application. Rochester was proposing filling in part of that Inner Loop, replacing it with a city street and using the surplus land for housing, bike lanes, and sidewalks. Beth loved it. "It's hard to look at the built environment and say, it should be totally different and this is how we're going to get ourselves there."

The project didn't get funded the first application round, nor the second. "It took a lot of time for us to get our arms around" the TIGER program, Beth says. And they had to spread their funding across regions and modes; Rochester had also applied for a grant to rebuild its Amtrak station. "You're trying to balance all these things." While Rochester wanted funding to fill in its Inner Loop, other cities applied for smaller grants simply to study the possibility of highway removal. One proposal suggested tearing down the Claiborne Expressway in New Orleans, the elevated highway that ripped through the Tremé neighborhood. The city received a $2.7 million TIGER grant and produced a slickly designed 138-page study that considered three scenarios for the bridge, including removal. Local reporters covered the study's publication. A bunch of local officials got three-ring binders. And then nothing happened. Similarly, when New York City studied the removal of the Sheridan Expressway in the Bronx, a well-designed report was produced and then quickly shelved. Those failures gave Beth an idea. What if the federal government created a dedicated program to tear down urban highways?

In March, when Biden announced his American Jobs Plan, it included $20 billion for a new program to "reconnect neighborhoods

cut off by historic investments and ensure new projects increase opportunity, advance racial equity and environmental justice, and promote affordable access," according to a fact sheet released by the
White House. "Too often, past transportation investments divided
communities—like the Claiborne Expressway in New Orleans or
I-81 in Syracuse—or it left out the people most in need of affordable
transportation options." Beth's fingerprints were all over the Reconnecting Communities program; she had spent months lobbying behind the scenes to get the idea in front of the president.

Across the country, freeway fighters celebrated this announce-
ment, a stunning admission of the damage wrought by urban highways on communities of color. In New Orleans, Amy Stelly's phone
started ringing minutes after the White House published its fact
sheet; with Biden's callout, the Claiborne Expressway became the
poster child of the problematic urban highway. Amy talked to *The
New York Times*, *The Washington Post*, NBC. She had been fighting to
tear down the Claiborne for years. Maybe, finally, it would actually
happen.

But as spring turned into summer and the so-called bipartisan infrastructure deal progressed through the House and the Senate, that
$20 billion was whittled down to $3 billion in the House bill and then
$1 billion in the Senate version. In Duluth, Minnesota—thirteen
hundred miles north of Austin along I-35—Jordan van der Hagen
watched as the proposed funding dwindled. A landscape architect,
van der Hagen had founded the Duluth Waterfront Collective, a
campaign to remove I-35 through Duluth's downtown and build a
boulevard in its place, reconnecting the city with the Lake Superior
waterfront. His vision was almost identical to Adam Greenfield's in
Austin: By removing the fourteen-lane highway and rebuilding a six-
lane urban boulevard in its place, van der Hagen said, the city could
reclaim more than twenty acres of land.

When he saw the $1 billion included in the Senate bill for Recon-
necting Communities projects, he sent an email to a listserv of several hundred freeway fighters across the country. "Does anyone else

think it would be impactful in any way to put together a letter voicing our support for an expanded reconnecting communities program?" he wrote. A billion dollars was barely enough to fund a few feasibility studies of highway removals, much less the actual teardowns.

Adam Greenfield was the first to reply. "Great suggestion," he wrote. "It would be good to start flexing our muscles collectively and this is the perfect opportunity to do so." Since he started Rethink35, Adam had been trying to connect with people fighting highway expansions across the country. He wasn't the only one. Ben Crowther had taken over Congress for the New Urbanism's Highways to Boulevards program in 2018. "Every time I talked to a group somewhere, they'd say, who else is out there like us?" he says. "Eventually I just started connecting them one-on-one. After each of those conversations, they'd come back to me and say, you know, it'd be great if we could just connect with everyone out there."

In the spring of 2021, Crowther organized a virtual "freeway fighters" social hour. More than ninety people signed on to the call; it took an hour just to get through introductions. Until then, Adam Greenfield didn't know about Jordan van der Hagen's campaign in Duluth— that on opposite sides of the same interstate, they'd conjured nearly identical visions. The call revealed the movement to itself, Crowther says. "By connecting groups fighting against expansions and groups advocating for reconnecting communities and seeing that their goals were pretty much the same, there were numbers there. Those numbers said, this is a movement." The group started meeting monthly, hashing out strategy and sharing updates from across the country. They talked about traffic modeling and the NEPA process, shared studies on air pollution and induced demand, reviewed federal legislation and lobbying efforts.

As Congress hammered out the final details of the Infrastructure Investment and Jobs Act, the Freeway Fighters Network sent a letter to leaders in the House and Senate. "On behalf of a nationwide coalition of campaigns seeking to reimagine freeway corridors in

American cities, we, the undersigned 96 organizations, call on our leaders in Congress to support the proposed Reconnecting Communities program," it began. "Funding the program at 1/20th of its proposed budget would greatly reduce this opportunity to reconnect and heal communities divided by interstate highways and other infrastructure."

On November 15, four days after I meet Beth for lunch, Biden holds a signing ceremony on the South Lawn of the White House. "This law makes this the most significant investment in roads and bridges in the past seventy years," he says. "So, what—what that means is you're going to be safer, and you're going to get there faster, and we're going to have a whole hell of a lot less pollution in the air." The Infrastructure Investment and Jobs Act included a five-year reauthorization of the surface transportation program, but with twice as much money—$643 billion—as the previous authorization. Of that $643 billion, $432 billion would flow directly to conventional highway programs, resulting in a 90 percent increase in highway funding. Meanwhile, $109 billion would go to transit. The federal government would spend only $1 billion to reconnect communities divided by highways.

10

HEMISPHERES

Dallas
I-345

Foam boards lean on metal easels arranged around the perimeter of a meeting room at St. Philip's School and Community Center in South Dallas. Printouts drape over fold-out tables, maps showing highlighter-yellow cross sections of I-345 superimposed over a gray street grid. Men and women in their thirties and forties, wearing khakis and magnetic name tags affixed to polo shirts, mill around the foam boards and folding tables. Tonight, TxDOT is presenting five options for the future of I-345, including removal—the first time the Dallas district has ever presented the full removal of an interstate highway to the public as a formal project alternative.

The schematic for the removal option shows the footprint of the highway, a wide orange swath the shape of an open parenthesis, cutting through a yellow street grid. The orange space, the land currently occupied by the highway, is labeled "new development opportunity." The so-called boulevard option is actually many boulevards, crisscrossing within the former footprint of the highway. Cesar Chavez Boulevard meanders through the orange land in two channels of sinuous gray.

That a removal option exists at all is remarkable. A decade earlier, Patrick Kennedy was some guy walking to work under an aging bridge. Now the Texas Department of Transportation is considering his idea to tear down an interstate highway. But even tonight, at the beginning of this formal consideration, it's clear to Patrick that TxDOT is not taking the removal option seriously. For one, TxDOT's redesigned street grid "looks as if it was designed and planned to maximize travel delay," Patrick tells me. To replace the car capacity lost from the highway, streets should run north-south, he says. Instead, they squiggle and swerve, dumping capacity east and west. "Somebody more cynical than me would say, are they presenting plans that they know have poison pills in them?"

I meet Ken Smith in front of a poster board showing a map of downtown Dallas slashed with ominous-looking red lines under the words WHERE DOES THE TRAFFIC GO? Ken, who is tall and lanky, with the build of a former basketball player, grew up in South Dallas. After retiring as the director of corporate communications for Blue Cross and Blue Shield of Texas, he moved back into his family's home on Bradshaw Street and started a nonprofit called Revitalize South Dallas. He's talking to Ceason Clemens, the deputy engineer for TxDOT's Dallas district. Ceason is the archetype of an engineer—visibly nervous in this free-form setting, clear on the facts but not necessarily on how to communicate them. She grew up in Plano, a suburb north of Dallas, the daughter of a mechanical engineer. She loved engineering, but she could never quite wrap her head around what it was that her dad did every day. When she took a class in civil engineering at Texas A&M, a lightbulb went on: This was something everyone could understand. "A lot of people can relate to what we do," she says. When her parents introduce her, they joke, "If you're ever stuck in traffic, it's because of Ceason."

To evaluate its five alternatives, TxDOT has modeled the future. Engineers studied current-day travel patterns on I-345 using anonymized cell-phone data to plot where car trips start and end. Using that data, TxDOT has predicted the future travel demand on I-345 in 2045 and then mapped that demand onto each of the five alternatives. On maps of downtown Dallas, which include the highway and surrounding city streets, a red line represents an increase in traffic—the thicker the line, the greater the delay. "High-level modeling shows with the removal we are adding nineteen thousand hours every weekday to the overall congestion in the subarea," Ceason says.

"Say that again?" Ken asks.

"What this board is trying to display is that with the removal, of changing it to a boulevard, we're adding nineteen thousand hours of congestion delay for every weekday as compared to today."

"Removal adds more congestion, more time," Ken says.

"Yes, sir," Ceason replies.

"So that's the worst-case scenario for people who are traveling from the south to the north," he says. His voice is low in pitch, musical; he talks slowly, deliberately. "Correct me if I'm wrong: Seventy-five percent of the traffic on I-345 comes from the south?"

That is correct, Ceason says. According to TxDOT's analysis of cell-phone data, most of the drivers who use I-345 start in South Dallas and end up somewhere in North Dallas, reversing their trip hours later. "Keep in mind, 180,000 vehicles travel through this corridor every day," Ceason says.

"So then the 75 percent of the cars that are using I-345 originating in the south, going to the north, they would be disadvantaged by the removal, disadvantaged being defined as spending more time on the road?" Ken asks.

"There's going to be a significant impact," Ceason says.

"That's what I'm afraid of," Ken replies.

The story of Dallas is the story of two unequal hemispheres: north and south. Mostly, this disparity is due to racism. In 1916, Dallas became the first city in Texas to legalize housing segregation. In the 1920s, the Dallas chapter of the Ku Klux Klan reportedly counted 13,000 members in a city with a population of only 160,000—one of every three eligible white men was a Klan member. In 1937, when an appraiser from the Home Owners' Loan Corporation visited Dallas, white neighborhoods in the northern crescent of the city were given the highest ratings, while most southern neighborhoods were shaded in red and yellow.

For decades, affluence headed north. The Central Expressway enabled the growth of suburbs like Richardson and Plano, even as it displaced hundreds of Black families, dispersing them across a still-segregated city. By 1950, Black Dallasites faced a housing crisis. Over the previous decade, builders had constructed sixty thousand homes within city limits. Only one thousand were available for Black families. "There is no room in the city for us. It is not right to wreck our homes when we have no place to go," said one homeowner displaced by the Central Expressway when it tore through North Dallas. Many families headed south, to lower-income neighborhoods like Oak Cliff and South Dallas. In 1950, a series of bombings tore through South Dallas, targeting eleven Black homeowners who had dared to move into the majority-white neighborhood. No one was injured, but no one was arrested, either. Dallas had an uneasy relationship with white supremacy, and it was best not to look at the matter too closely.

Since the 1950s, Dallas has been divided by Interstate 30, which cuts across the city east to west, underlining downtown. Almost any way you can measure well-being splits along this line. North of I-30, the average life expectancy is twenty-two years longer than south of it. Southern Dallas accounts for 64 percent of the landmass in the city of Dallas but only 10 percent of the city's assessed property value. The area has lost 17 percent of its jobs since 2000, even as the Dallas–Fort Worth region led the nation in overall job growth; most of those jobs are located in northern Dallas and in the booming suburbs beyond. Most people who live in southern Dallas commute north to go to work. On a heat map of average commute times, the north-south bifurcation holds: People in southern Dallas commute farther and longer. In 2018, as part of a fair housing analysis, the city mapped household location by race. The map had an uncanny symmetry, with two triangles—one Black, one white—reflected away from downtown. Hispanic households filled in on the east and west. "In Dallas, the spatial segregation of the Black and Hispanic populations has a much more significant impact than poverty on the populations' access to opportunities," the report concluded. It wasn't that Dallas residents were explicitly being denied access to opportunity, the report seemed to say; they just couldn't get to the places that opportunity existed.

"You have to remedy what the highway has created," says Terry Flowers, the principal at St. Philip's School, after Ken leaves and Ceason and I continue to hover around the foam boards. Terry lives in DeSoto, a suburb south of Dallas. When I-345 was built, it created a chain reaction, he says. Neighborhoods could remain segregated because the highway did the work of integration. White-owned companies employed Black and Hispanic workers, and then those workers went home at the end of the day, to a community somewhere else with someone else's problems. "That chain reaction still exists today. We haven't done anything about it," he says. "Now you are going to add insult to injury to make it even more difficult to access your livelihood without addressing all of those other circumstances that were generated as a result of that decision."

The highway did not create segregation, but it reinforced it, siphoning opportunity away from Black neighborhoods. The cost of admission to that opportunity was a car. For Flowers, removing the highway without bringing that opportunity back to South Dallas was like cutting off the community's supply of oxygen. Sure, a highway-free future would be great. But people needed to breathe today.

The next morning, I drive out to TxDOT's Dallas district office, a sprawling complex in Mesquite, a suburb just east of the city. I'm escorted into a room filled almost wall to wall with a massive mahogany conference table. In addition to Mo Bur, who runs the Dallas

district, and Ceason, whom I'd met the night before, there are two
public information officers and two bonus engineers. I'm here because I want to know how TxDOT will decide the future of I-345—
how will the agency pick from the five options it has presented to
the public? The party line is that TxDOT is agnostic to what happens
to the highway—whether it is torn down or expanded, elevated or
buried. The decision will be made after the public weighs in. Everything happens in the passive voice: Agencies are engaged; stakeholders are consulted. Elected officials will be "visited with," while public
feedback "is considered."

"I think a lot of people sometimes think this is an election," Mo
says. "And if this option gets all the votes, then that's what TxDOT is
going to do. This is not an election. It's not the option that gets the
most votes."

"Why not?" I ask. "It's public money."

"It's public money, but we're a state agency. We don't create policy," Mo says. "We follow the rules that are set by the state legislature, the governor of Texas. We have a Texas Transportation
Commission that sets policy and rules and guidelines."

I ask about the projected nineteen thousand hours of traffic delay.
"What does that mean in minutes for an individual driver?"

They don't know yet; the traffic model has not been refined to this
level of specificity. "It's not an easy math equation," Ceason says.
The delay will vary based on time of day—rush hour versus midnight—as well as origin and destination.

But how can anyone weigh in on removing the highway before
they know whether their daily commute will increase by five minutes or thirty? I ask. "Seeing nineteen thousand hours, I would imagine many people say, oh, no, we don't want that option." But what if
those nineteen thousand hours represented only a few minutes of
delay for each driver? In that case, someone could, in theory, decide
the delay was worth it for all the other benefits that might come
from removal—everything that might be built in the highway's

place. It was a trade-off—this for that—but TxDOT's presentation was all cost, no benefit.

"We're not there yet," Mo says. "This is still a bigger picture." Over the coming year, Mo and Ceason will sift through thousands of public comments—they will synthesize conversations with elected officials and other stakeholders—and then they will narrow down the alternatives, refining one or two to present back to the public. "We're going to present, hey, here's what we heard from the community," Mo says. "Here's what we think the corridor should be."

Traffic models are slippery things, presented as scientific fact when they are actually best guesses, extrapolations of the future from the past. "A lot of people want to avoid the conversation of what is desirable, and they just say, 'Well, the number says this, and so we've got to accommodate it,'" says Robert Goodspeed, a professor of urban planning at the University of Michigan. He calls this "colonizing the future." If you plan for a future of car-centric sprawl, that's what you'll get. And TxDOT's models do not consider the trip not taken; they do not acknowledge what happened in San Francisco and New York City and Milwaukee, when highway capacity was removed and people simply drove less. "I think traffic engineers tend to think traffic is like a liquid. If the pipes aren't big enough, then it gets plugged up and overflows. The solution is building bigger pipes," Goodspeed says. "But all of the evidence says that that's not true, that instead [traffic] is much more like a gas, meaning the volume of traffic congestion will expand to take up the capacity allowed." And if traffic can expand, it can also contract.

Finally, I ask the engineers to talk about engineering. "How do you remove a highway?" I ask. "Like, physically? Let's say that tomorrow everyone in Dallas said, let's get rid of I-345. How would you do it?"

There's a pause. "Very carefully," Mo says. Everyone laughs. "It's dynamite. It would be blowing stuff up left and right." Ceason jumps in; TxDOT will not be blowing up downtown Dallas. "Typically we put a hammer on it. Start hammering on it and drop it," she says.

"So the deck is concrete," Mo says. "All the beams and all the gird-
ers underneath are steel. So it would be socketing those with torches
and removing them and then the columns . . ." He trails off and looks
at Ceason. "Usually they hammer every bit of it," Ceason says. "On
a giant excavator, they'll put a hydraulic hammer on it. You eventually get to a weak point and it comes down."

"So you're hammering the deck, and then you somehow have to
get through the steel that's below it?" I ask.

"Sometimes they'll cut it with torches; sometimes they'll use
munchers and actually munch up the steel and haul it off," Ceason
says. "Then you turn to the columns; they are made out of concrete
reinforced with steel, the rebars."

"You go from the top down," Mo says.

"How long would that take?" I ask.

Mo and Ceason look at each other and shrug. Six months, maybe?
It's pretty easy to tear down a highway, once you decide to do it.

Lockridge Wilson drives on I-345 almost every day. Some days, his
commute from his home in South Dallas to his job in North Dallas
stretches close to an hour. Lockridge works as a maintenance supervisor at an apartment complex off Lemmon Avenue in Highland
Park. To get there from his house off Pine Street in the heart of South
Dallas, he often heads north on I-45, the highway rising over Deep
Ellum as it sweeps into downtown and becomes—imperceptibly—
I-345. He can't imagine what it would be like to drive to work without the elevated highway. "Everybody who lives in the south has to
go north to work," he says. "In the morning, it's traffic under the
freeway and on top of the freeway. If they take the overhead down,
they going to have to add more lanes."

When Lockridge was born in 1955, his parents lived in Rhoads
Terrace, a sprawling public housing complex of redbrick apartments
a few blocks from the South Central Expressway. When Lockridge
was a kid, after his dad was able to get a federally backed mortgage,

the family moved into a house near Lincoln High School, still just a few blocks from the highway, which had spliced apart streets and split the neighborhood in half. There was a corner store just across the highway from Lockridge's house, but the closest intersection was three blocks south. "You had to walk all the way down to there to get on the other side. Or you take a chance running across," he says. He took the chance, every time. Starting when he was eleven or twelve years old, Lockridge would walk north on the service road that ran alongside the Central Expressway up to the Forest Theater, the regal structure looming over Forest Avenue. "You got three features a day: You got a movie, you got a cartoon, and you got a western," he remembers. "The bus was eleven cents, but you don't want to spend your money until you got to the movies, so we used to walk."

The Forest Theater had survived the construction of two highways running within a block of its front doors. When the South Central Expressway tore through South Dallas, it cut a swath directly to the east of the theater. When Interstate 45 came through the neighborhood in 1970—connecting Houston to Dallas—planners routed it directly to the west of the Forest Theater, stranding a pizza-slice-shaped sliver of South Dallas between the two elevated expressways. For decades, the theater's white marquee faced this chaotic convergence of cars. Rising over the marquee was a tower of green, spelling out the word "Forest" and topped by a red globe—a dropped pin, visible throughout the neighborhood.

In the 1950s, the Forest Theater was one of the only places with air-conditioning in South Dallas. Lockridge was still a kid when Ike and Tina Turner performed there, when Gladys Knight & the Pips came through. The Forest was on the national jazz and blues circuit; B. B. King and Little Milton both performed on its stage. But Lockridge didn't go to the Forest Theater much once he learned how to "shoot hooky" from school. He was incarcerated for the first time when he was sixteen years old and spent most of the next decade in prison. In 1980, when he was twenty-six, he decided to change his life. He got out of prison, got married, and got a job doing maintenance for the Dallas Housing Authority. He had two daughters and two sons and bought a house in the neighborhood he'd grown up in. By the time his youngest son, Lakeem, was born in 1991, the Forest Theater had long been abandoned in the armpit between Central Expressway and I-45, which merged just before sweeping north into downtown Dallas and lifting above the city to become I-345.

One morning in 2017, a woman named Elizabeth Wattley got a call from a white couple named Linda and Jon Halbert. The Halberts had bought the Forest Theater after seeing a FOR SALE sign on the marquee while driving south on the highway. They wanted to restore the historic building and reopen its doors as a performing arts and community center serving South Dallas. "The biggest threat to the ongoing viability of that theater is a one-word answer: poverty," Jon Halbert told *D Magazine* in 2017. "It reflects what has happened in South Dallas and Fair Park in forty years, a history of disinvestment. The theater reflected that economic reality."

Elizabeth, who is Black, was born and raised in Cedar Crest, a sprawling neighborhood stretching south of downtown along the western edge of I-45. "South Dallas has always been a big part of my life, kind of like my backyard," Elizabeth says. She learned how to swim at Exline Recreation Center in the heart of South Dallas, taught by a local teenager named Erica Wright. After earning a bachelor's degree in economics from Spelman College in Atlanta, she'd moved back to Dallas and started working as the director of service learning at Paul Quinn College, the state's oldest historically Black college. When Michael Sorrell took over as president, the struggling college couldn't afford its football program. So he committed Texas heresy and tore up the field, converting it into an organic farm. Under Elizabeth's watch, students planted peas, lettuce, and carrots from goalpost to goalpost. "I was that person in 2010, harvesting radishes and really figuring things out," she says. "That's when I figured out that you can solve social problems with business principles."

The Halberts invited Elizabeth to come see the theater. On the interior walls, chunks of plaster had dissolved, cracks spreading like spiderwebs across the vaulted ceiling. A layer of dirt and dust coated the black-and-white-checkered floor. Past the restrooms—"ladies" and "men" marked with cursive no-longer-neon lights—was an art deco lobby painted a deep carmine red, spiraling upstairs to the balcony. There were no seats, just a stepped terrace sloping toward the stage. Elizabeth was overwhelmed. South Dallas had been neglected by the city for generations; there were miles of crumbling clay pipes below the theater that would need to be replaced before a single toilet could be flushed. And she knew how many people had tried—and failed—to bring the historic theater back to its former grandeur.

After high school, Elizabeth's swim teacher had changed her name to Erykah Badu, and, after releasing two platinum records and winning a few Grammy Awards, she decided to lease the theater and bring music back to South Dallas. For years, Badu fought to make the Forest Theater profitable, bringing in artists like Prince, Snoop Dogg, and the Roots. She started an after-school program for kids from the neighborhood, teaching them how to sing and dance, just as she had been taught when she was young. But the theater was too expensive to maintain. "Financially, it's a burden," she said in 2008, shortly before she gave up the theater's lease. "It's a big job to take on without having the correct kind of sponsorship and backing."

When Wattley agreed to take over the Forest Theater, she knew she needed to do something different. She started by holding a series of meetings with South Dallas residents. What did they want to see happen in the theater? What did the neighborhood need? In 2017, she propped a large chalkboard across the theater's front doors, with a sign inviting people to leave notes and ideas. Artist space, bakeshop for kids, supper club, they wrote. One morning, Elizabeth arrived at the theater to find many of the replies had been erased by someone who had scrawled, simply, "Apollo Theater."

In November 2021, Forest Forward—the new nonprofit that would own and manage the theater—announced a plan to raise

$75 million to restore the building and reopen it as a twelve-hundred-seat music venue and event space. The reimagined Forest Theater would be for the community that contained it: It would include a coffee shop and rooftop patio, open to all, as well as a thirteen-thousand-square-foot arts education center that could be used by students at a nearby school. "What an opportunity we have before us," Elizabeth says. "What an opportunity to break the cycle of broken promises to the South Dallas community, what an opportunity to come together and reignite the heartbeat and the soul of this neighborhood."

Nine months after launching the campaign, on a hot morning in September, Elizabeth stands before a roomful of potential funders, seated in rows of rented event chairs. Two open air ducts snaking across the floor blow cold air on the front rows, rippling the white tablecloths covering the bar tables on the perimeter. Behind Elizabeth, on the crumbling stage, red velvet curtains heavy with dust ripple in the artificial breeze. This is Elizabeth's pitch to funders: help the theater reopen its doors to the public and help bring wealth and opportunity back to South Dallas, restoring the jobs and homes the neighborhood lost when it was carved up by highways.

For Elizabeth, the restoration of the Forest Theater was "the big sexy thing" that would drive the nonprofit's real work: neighborhood revitalization. Only 25 percent of residents of South Dallas owned their own homes. The median income hovered around $25,000. And yet after decades of disinvestment, gentrification had started to creep into South Dallas. When the city announced a plan to redevelop Fair Park, the site of the state fair, real estate developers started eyeing the neighborhood just to its west. Some blocks in South Dallas saw property values increase by more than 100 percent one year to the next, as rows of $60,000 homes were swept up by investors. "You want new development," Elizabeth says. "But you want to keep people. You don't want to displace residents." If successful, the restoration of the theater would bring wealth and development to the neighborhood around it. What Elizabeth wanted to do was direct that wealth to the people who needed it most.

The theater was surrounded by acres of vacant land and abandoned houses. Elizabeth started quietly buying properties, hoping to assemble enough land to build 150 units of mixed-income housing on four acres just south of the theater. She talked to leaders at Dallas Independent School District and met with Romikianta Sneed, who was then the assistant principal of the Martin Luther King Jr. Learning Center, a struggling elementary school a block south of the theater. Elizabeth saw the school as an anchor for the theater's restoration work. "The school will be the reason families move to the neighborhood," she says. "That will be the reason for the diversification of the socioeconomics and the people—because people want to attend that school so that their kids can get a good education." In 2020, the school reopened as the MLK Arts Academy, with a new focus on arts education, including visual arts, music, and dance. Sneed became the school's principal. Existing students could stay at the school, and nearby families were given priority admission as the arts academy added two grades, expanding its new middle school into a building funded by a voter-approved bond as well as Forest Forward. In the future, Elizabeth imagined, students could take classes at the theater; they could learn graphic design and podcast production and film editing. Elizabeth started working to acquire the right-of-way to a block of Harwood Street, which connected the theater to the school, hoping to eventually close the street to traffic and create a pedestrian pathway between the theater and the school. "The theater alone would have never made it," Elizabeth says. "It needs to be intertwined to sustain and stay up."

After Elizabeth's presentation, the future funders wander through the crumbling theater, walking upstairs and peering out a balcony that overlooks the parking lot. Just beyond the cracked asphalt stretches the South Central Expressway, which was renamed S. M. Wright Freeway in 1995 in honor of a local pastor and civil rights leader. The highway splits from I-45 just south of downtown, dipping under Martin Luther King Jr. Boulevard—as Forest Avenue had been renamed—before sloping gently upward. A block south of the theater, a blinking arrow directs drivers onto the frontage road. In front of the blinking arrow, protected by orange traffic barrels, a sign reads, FREEWAY ENDS.

For nearly two years, TxDOT has been working to remove the two-mile stretch of S. M. Wright. Three miles south of downtown, S. M. Wright swerved abruptly east as it joined U.S. Highway 175, creating a hairpin turn so dangerous that people started calling it Dead Man's Curve. In the late 1990s, planners suggested a new road that would stretch northwest over the floodplain of the Trinity River, starting at Dead Man's Curve and running alongside the river for eight miles until it joined I-35E northwest of downtown. As it advanced plans for the Trinity Parkway, TxDOT announced that the new road would link U.S. 175 with Interstate 45, avoiding the sharp change in direction at Dead Man's Curve and obviating the need for the two-mile stretch of highway between the new interchange and downtown. In 2010, the state held a public meeting in South Dallas and announced that it intended to remove the elevated highway and rebuild the road as a tree-lined boulevard.

But as residents learned, TxDOT's version of a boulevard was a wide, fast, six-lane road. South Dallas residents pressed for a narrower roadway. "There was a lot of harm done to this community when those roads came in," a community organizer named Henry Lawson told The Dallas Morning News in 2010. "Now, you say, that was thirty to forty years ago. But don't you have some obligation, some responsibility—a social responsibility—to help us make our neighborhood whole again?" He wanted TxDOT to rebuild Central Expressway as a four-lane street, turning the excess land over to a community land trust that could help residents open shops and restaurants with low-cost leases. TxDOT was unwilling to consider the idea. "If we build a project with just four lanes, we know we're basically building a $40 million parking lot," said Tim Nesbitt, the engineer in charge of the project. Besides, he said, "TxDOT is not in the development business." It was in the traffic business.

But while many residents were disappointed with TxDOT's plans, just as many were thrilled; a six-lane parkway was a whole lot better than an elevated highway. On the new S. M. Wright Parkway, there would be stoplights every few blocks, a narrow park in the median, and dedicated walking and biking trails on either side. At the corner of Martin Luther King Jr. Boulevard, TxDOT renderings showed people walking west across the parkway, steps from the front doors of the Forest Theater.

The Trinity Parkway toll road was a boondoggle from the get-go. The river was a wide channel that surged with fast-flowing water during rainstorms, protecting the city from flooding. The road would pave over this fragile habitat. In 2015, a federal study found that the toll road would *increase* traffic on nearby highways. Two years later, after a decade of bitter debate, Dallas's city council officially rejected plans to build a highway over the city's floodplain. But by then, TxDOT was years into planning the removal of a short stretch of the South Central Expressway. It couldn't very well abandon its plan to fix Dead Man's Curve—the name spoke for itself. So in 2020, TxDOT began tearing down the first urban highway in the state of Texas.

11

UNCERTAINTY

Houston
I-45

Houston in the summer is lush and buzzing, sweat soaked and full of life. The terra-cotta pots assembled on the front porch of the last house on Ishmeal Street overflow with flowers and escaping vines. In the front yard, an old well—a round brick cylinder covered by a corrugated steel roof—brims with zinnias. Heavy purple hydrangeas lean over a cinder-block planter. Sunk into the grass in front of a hedgerow, a bright red yard sign reads, FREEWAY—NO WAY!

Molly Cook pulls up in her chrome minivan—the only car that will fit her harp, which she still plays—and gets out carrying clipboards and a stack of postcards. She greets Elda and Jesus Reyes, the couple who live in this house next to the highway. On the eastern edge of their property is an unpaved drainage ditch, six feet across. Beyond that is a fence and then the back side of a strip mall, a row of parking spots, and then, finally, I-45. Molly gives a clipboard and stack of postcards to Elda and Jesus and the three Stop TxDOT I-45 volunteers who have come out on this sweltering Saturday to canvass the neighborhood. Elda brings out cold water bottles from inside the house, the condensation dripping onto her jeans.

Although it had promised to "expedite its efforts," the Federal Highway Administration's pause on the North Houston Highway

Improvement Project had dragged on throughout the summer. A few weeks earlier, at the Texas Transportation Commission's monthly meeting, J. Bruce Bugg, Jr., threatened to pull funding for the entire project from TxDOT's ten-year plan. The projected cost of the NHHIP had increased to $9 billion, which represented 12 percent of the state's total funding for transportation projects, Bugg reported at the commission's meeting. He was disappointed that 12 percent of the state's transportation budget was being held hostage by the Federal Highway Administration. "This is absolutely unprecedented that the Texas Department of Transportation is facing a situation like this on as major of a project and impactful a project. . . . We're about to break ground and then we find ourselves being halted right literally in our tracks," he said. He proposed putting the project back out for public comment, offering a take-it-or-leave-it proposition: accept the project as it had been designed or remove the project from TxDOT's ten-year plan and allocate the discretionary funding elsewhere in the state.

This is a sham survey, Molly says now. It presents a false choice, forcing Houston residents to accept a highway widening or risk losing funding for safety and drainage and transit access. All week, Stop TxDOT I-45 had been knocking on doors across Houston, asking people to fill out the poll TxDOT released on SurveyMonkey. "Instead of working with the community, TxDOT wants to just take their toys and go home," she says. "So we're calling their bluff, telling them to remove the funding until they can come up with a better project."

Molly heads a block north to Red Ripple Road as Elda and Jesus begin to work their way down Ishmeal. "Buenos días," Elda says, whenever someone opens a door—assuming, correctly, that everyone on their street speaks Spanish. Most people on Ishmeal have never heard of the highway expansion. Largely, TxDOT's outreach has been in English, which many people in this neighborhood—including Elda—don't speak. And why would anyone pay attention? The project had been in the works for a decade. In the meantime, Houstonians had been hit by a hurricane and a devastating winter storm; people had more urgent disasters to fight. Still, Jesus is frustrated that so few of his neighbors know about the project. "It doesn't affect them," Jesus says.

Elda grew up in a small town in the Mexican state of Tamaulipas, a couple hours south of the border. She moved to Houston after she married Jesus, whom she met through a friend. Was it love at first sight? I ask. "For him," she quips. She had never imagined herself living in such a big city, driving on its fast, tangled highways. Jesus worked as a carpenter, and Elda stayed home to raise their three kids. After a decade moving from apartment to apartment, in 2004 they got a realtor and started looking to buy a house. When they first pulled up outside the house on Ishmeal Street, Elda looked across the street at a used-car lot, hidden behind a corrugated metal fence. She looked at the small house. She didn't want to get out of the car. Jesus persuaded her to go inside. She pointed out the small kitchen, the one bathroom the five of them would have to share. Jesus pointed out the big yard. He said, "We'll fix it up. We'll make it ours." As for the impound lot across the street, well, at least they wouldn't have problems with their neighbors. The Northside, a majority-Hispanic community, was affordable and quiet. There wasn't much traffic on the dead-end street, so their kids could play outside. And they could afford the monthly payments—just over $500 a month—on Jesus's salary alone.

"Honestly, it wasn't the house of my dreams," Elda says. But it was a start. On nights and weekends, Jesus tore out the tiny kitchen cabinets and built new ones, varnished in rich chestnut. He redid the baseboards and installed new crown molding, intricately patterned. They installed new doors, new windows. They scoured Houston for furniture and hung framed family photos on the walls. "You start falling in love when you add little touches; you start liking it more and more," Elda says. They planted fruit trees in their backyard—peach and plum and avocado—and waited for them to flower.

In the spring of 2019, Elda and Jesus paid off their mortgage. The house was theirs, free and clear. Three months later, Elda arrived home from work, checked the mail, and found a pamphlet from the Texas Department of Transportation. It was printed in Spanish as well as English, so she sat down to read it. "The Texas Department of Transportation is exploring and refining alternatives to address the continued growth facing the Houston area," it began. Since 2011, the agency had held many public meetings on the North Houston Highway Improvement Project. This was the first Elda had heard of it. She kept reading. "To construct NHHIP, TxDOT would acquire approximately 450 acres of new right-of-way. . . . TxDOT is already working to acquire some properties and in the coming months will be rolling out a project-specific relocation assistance program." Elda turned on her computer and found the right-of-way maps online. The proposed footprint of the expansion was rendered in translucent red, a watercolor wash over the city. The right-of-way line ran directly through their home.

Elda was devastated. "Many people don't understand because they say, 'Well, that's an old house.' Well, yes, but we've all lived here, it's our first house." She didn't want to move. Her family's memories were contained in this house. It was where her kids had grown up—"where they had become themselves." And where would they go? Houses had become so expensive in Houston. "It's ugly. Now I understand people who, for example, lose their homes in a fire. I understand that feeling of losing everything," Elda says. "Like how a hurricane takes everything with it. It's not the value . . . it's the sentimental value."

Now, as they walk slowly down their street, Elda feels slightly more hopeful. So many people are affected, she insists. Surely they won't go through with this. Elda and Jesus have never canvassed before—never knocked on strangers' doors to ask for something. But if their neighbors won't help them, who will?

Midway down the block, a middle-aged man invites Elda and Jesus into his home before he hears why they are at his doorstep. "It's so hot outside, come in, come in," he says. He pulls out seats and offers cold bottled water before asking, in Spanish, "What can I help you with?"

"We live down at the end of Ishmeal," Elda says. The man sits in a swiveling office chair. "We're directly affected by the I-45 expansion. We're going door to door to try to get signatures, but through an online survey." The man sees the QR code on the flyer she's just handed him and goes to get his smartphone. "I wasn't sure when that was going to happen," he calls from the other room. He'd heard about the project years ago but lost track of it.

"It's on a pause," Jesus says. "We're trying to get people who support us to see if we can stop it permanently." The man returns, nodding. "Many people support it because they say the traffic is bad and they want to fix traffic," Elda says. "But if you expand the highway, it will just fill up with more cars. And if there are more cars on the highway, there will be more pollution." The man nods. He, too, understands the phenomenon of induced demand. "They are going to displace other people. We don't know where we're going to go. Home prices are so high," Elda says. The man pulls up the survey on his phone. He thanks Elda and Jesus for stopping by. "Good luck," he says.

Meanwhile, a block north on Red Ripple Road, Molly knocks on the door of a low-slung redbrick house. Someone who lives at this house loves to garden; potted ferns and hanging pothos frame the porch, which is shady and cool. A Black man in his sixties answers the door, his face a question mark. Molly jumps right in: "I'm with a group fighting the I-45 expansion. Have you heard about that?" she asks. The man, named Mike, shakes his head. "We're trying to stop it from happening, basically," Molly says. "It's supposed to take a thousand homes. There's a lot of public pushback."

"Where will it go?" Mike asks.

Molly points down his street; there, she says. "It's going to be much closer to you guys."

Ever the nurse, Molly has excellent bedside manner. She's direct and to the point, adept at swiftly explaining complicated procedures. She knocked on a hundred doors a day when Beto O'Rourke ran for Congress, and it shows: She is indefatigable, unintimidated. "Don't you think TxDOT should have to go door to door, just like we're doing, sweating on the streets?" she says to me, as we continue walking. A few doors down, Molly meets a guy in his thirties named Sebastian. He leans against his door frame, the air-conditioning rushing out into the heat. The expansion plan is news to him. The highway will get wider, Molly says, on both sides. "What are they going to do with Mattress Mack?" Sebastian exclaims, his first response. "That's the man around here. How are you going to shut him down? He takes care of the neighborhood."

Mattress Mack is Jim McIngvale, a seventy-one-year-old Houston celebrity and furniture salesman. Mattress Mack, famously, gives away mattresses when the Houston Astros win the World Series. Year-round, a sign outside Gallery Furniture announces, IF THE ASTROS WIN IT ALL, YOUR MATTRESS IS FREE. In 2017, when the Houston baseball team won the World Series, Mattress Mack refunded $10 million to customers who had bought mattresses during the playoff run.

Mattress Mack has stood behind the customer service desk at Gallery Furniture almost every day since 1981, when he and his wife opened the storefront on the frontage road. From this perch, he can see I-45. He has looked at the highway twelve hours a day, seven days a week, for forty years. He watched it when it flooded during Hurricane Harvey—when he opened up the store as a shelter and invited hundreds of people to sleep on display beds and couches—and when it froze over during Winter Storm Uri and he again opened the store to hundreds of people who had lost power. "We need extra lanes on that highway like I need a hole in my head," he tells me when I swing by the store on my way back downtown. "I think it's going to destroy the Northside for ten years and put a lot of people out of their homes."

Mattress Mack is a Republican. He is a small government guy, a longtime supporter of the Tea Party. In 2020, he was appointed by Governor Greg Abbott to serve on the Strike Force to Open Texas, which reopened the state for business during the COVID-19 pandemic. The highway expansion is "a big government boondoggle," he says. The expansion would consume a small sliver of his parking lot, but that's not what he finds offensive. "It's for the fat cats: the highway department and the contractors. Not the people on the Northside who are grinding out a living. These people are on the edge. They don't need more grief coming at them." A $9 billion highway was the last thing the government should be spending its money on. "It's a boondoggle for everyone. How much more money can we waste?" He'd heard about TxDOT's survey. "I guess the newest thing is an either-or," he says. "Either you do everything or you get nothing. That's a mafia pitch. An offer you can't refuse."

A few hours later, the heat has begun to subside as the patio of Axelrad Beer Garden fills with people wearing red T-shirts. A handmade paper sign hangs on a back wall, twenty-six red characters threaded on a string: HAPPY BIRTHDAY STOP TXDOT I45!

Tonight, the group is celebrating its second birthday. Elda and Jesus have rested and recovered and driven down to Midtown. When Modesti Cooper arrives, I offer to buy her a beer. "I gotta be careful," she says. The last Stop TxDOT I-45 event she went to turned into a pub crawl. This oddball group of people, thrown together to oppose a highway, stayed out until the bars closed. Tonight, there is a DJ, a silent auction—bike service, beer, books—and poster boards propped on easels recapping the last two years: WHY REMOVE THE FREEWAY? and STOP GOES TO AUSTIN: TXLEGE DAY 2021. The second poster includes a photo of fourteen people lined up in front of a fifteen-passenger van, bleary-eyed at 6:00 A.M. They were on their way to Austin to testify at the state capitol on several proposed bills, including a measure that would put to a public vote whether to allow state highway funds to be spent on things other than roadways. (None of the bills made it out of committee.)

Since Susan Graham made her yard signs and Molly started knocking on doors, Stop TxDOT I-45 had become a formidable grassroots organization, garnering the attention of local and statewide elected officials. The group hosted concerts in bayou parks that would be covered by highways. They painted a giant banner—NO I-45 EXPANSION—and hung it on the side of a freeway overpass during rush hour. They showed up to testify—at the city council, at the Houston-Galveston Area Council, at the Texas Transportation Commission. They organized social gatherings, just for fun. In December 2020, they organized a community block walk, traversing the footprint of the highway expansion, stopping at the homes and businesses that would be destroyed. They spray-painted stenciled letters onto sidewalks across the city: TXDOT WANTS A HIGHWAY HERE.

In some ways, Susan and Molly were an unlikely team; their age difference spans exactly thirty-four years and three days. Tonight, Susan wears a floor-length turquoise sundress, a yellow cardigan, and chandelier earrings. She is glamorous. Molly is wearing the same thing she's been wearing all day: a red Stop TxDOT I-45 T-shirt, jean shorts, and Converse tennis shoes, her red hair piled in a messy bun on the top of her head. But together, they have somehow made fighting freeways—there is no other word for it—*cool.* By 8:00 P.M., the patio is packed. Nearly sixty people have gathered on a Saturday night to support the movement to stop highway expansions in Houston. It's an unexpectedly young crowd, full of hipsters in their twenties, tattooed and stylish, wearing cutoff jean shorts and drinking IPAs.

Mark, Molly's dad, bought the cake for tonight's birthday party, a giant sheet cake covered in white frosting. A red border frames the message, and in the middle, written in looping cursive frosting, it says, "Stop TxDOT I-45 Happy Birthday!" "The people at the bakery had no idea what I was talking about," he says, laughing. "I finally just had to hand them a piece of paper and say, yes, this is what we want the cake to say."

A month later, Molly wakes up before dawn to drive to Austin for the Texas Transportation Commission's monthly meeting. The results of the survey were in. The commission would announce whether it intended to keep or remove the billions allocated to the North Houston Highway Improvement Project in its ten-year plan. The meeting room is unusually full, activists in red T-shirts sitting next to an increasingly cranky cohort of middle-aged white men, elected representatives who are used to getting their way without a fight. "We've had a number of people speak pro and con," Bugg says. "Two-thirds of the members of the public that participated in our public outreach voted in favor of this project, whereas a third voted against the project."

He does not mention how many people participated in this hastily assembled public survey—just over 8,000 in a city with a population of 2.3 million—but to him the mandate is clear: The project should remain. Still, there is an elephant in the room. "Until the FHWA reauthorizes TxDOT to start this project, TxDOT cannot do so on its own," he says. "I'm pleased to say the mayor of Houston as recently as just yesterday came out publicly in favor of this project." So he gives the Federal Highway Administration a deadline. "I believe we should give the Houston community ninety days to work with FHWA," he says. Ninety days for the federal government to wrap up its investigation and offer a viable path forward. If no such path emerges by the end of November, the commission will be forced to remove funding for the project.

Two weeks later, at a city council meeting, Houston's mayor, Sylvester Turner, corrects the record. He does *not* support the highway expansion, he says. He'd simply asked the commission to consider signing a memorandum of understanding, committing TxDOT to certain mitigation measures. "Chairman Bugg misrepresented my position totally," he said. "That is what makes a lot of people distrust TxDOT. Because to make a statement like that, a public statement of my support without signing the MOU, just raises a lot of flags and creates a great deal of distrust."

Bugg's ninety days come and go. On a drizzly Monday afternoon in December, fifty people gather in a windowless room in the historic St. John Missionary Baptist Church on Emancipation Avenue, just south of where I-69 and I-45 converge. They are community members, advocates, elected officials. They tug off jackets and sit in cold metal folding chairs tucked behind fold-out tables covered with plastic tablecloths. They pull down masks to eat sandwiches out of white cardboard boxes. Fans twirl on the low-slung ceiling. Congresswoman Sheila Jackson Lee, who represents Texas's Eighteenth District, which spans most of Houston, stands behind a lectern at the front of the room, framed by two plastic palm trees in wicker baskets. "We're here to tell the true story of the I-45 project," she says.

To the left of Jackson Lee sits Stephanie Pollack, the deputy administrator of the Federal Highway Administration. Pollack is in Houston along with several of her colleagues, including Nichole McWhorter, the head of FHWA's Title VI team. Nearly nine months after the department paused the North Houston Highway Improvement Project because of civil rights concerns, they've flown to Houston to investigate them. With only a few days' notice, Jackson Lee hastily assembled this community listening session so that Pollack could hear from people directly affected by the highway expansion. One by one, people stand up and tell Pollack how Houston's highways have affected their lives and communities—how their neighborhoods have been carved up, how they have been displaced, how they have been made sick by breathing polluted air. "We have been sliced and diced," says Joetta Stevenson, the president of the Greater Fifth Ward Super Neighborhood association, who happens also to be Curley Guidry's cousin. "Please do something. This has got to stop."

"I came to Houston today to listen and learn. I have learned a lot," Pollack says after everyone in the room has spoken. The federal agency was actively investigating Title VI complaints, she said. "Because we are in the middle of that investigation, because we are in the middle of an audit on how TxDOT has complied with the National Environmental Policy Act, we can't give you answers today," she says. "But I can tell you that we hear you. That what you're saying will be taken into account when we complete those investigations."

In a meeting earlier that day, Pollack had rejected the deadline Bugg gave FHWA, stressing that the investigation was not on an arbitrary timeline and that it would continue for as long as it needed to. "We're going to work to solve the problem," Jackson Lee says now, "but not with TxDOT telling us what to do."

The group filters out of the church and boards a METRO bus for a tour of places affected by the highway expansion. They visit Clayton Homes, a 296-unit public housing complex that would be scraped clean and paved over. TxDOT purchased the property in 2020, well before the federal government intervened, and started moving tenants out. By late 2021, it is sparsely occupied. They stop at Bruce Elementary School, which occupies a redbrick building just below the soaring spaghetti junction of I-10 and I-69. The expansion would bring the highway even closer to the school. Instead of being relocated, students—who already suffered from asthma at twice the rate as other students in Houston Independent School District—would be forced to breathe in noxious fumes from traffic brought 150 feet closer to their classrooms and playground. In 2019, the nonprofit Air Alliance Houston modeled air emissions at the school and found that bringing the highway even closer could cause substantial increases in traffic-related air pollutants like benzene, a chemical compound known to cause cancer.

The tour continues north to a neighborhood called Independence Heights, incorporated in 1915 as the first Black city in Texas. In the 1950s, the construction of the 610 Loop spliced the community in half, wiping out 250 homes and bisecting White Oak Bayou, one of dozens of creeks that functioned as a natural drainage system for the city. When the Texas Highway Department built the interchange at I-45 and 610, it constructed a culvert for the water to flow through. But the culvert was too small. It got clogged. Water backed up. And so, Independence Heights—still a majority-Black neighborhood—often flooded. "We're like a swimming pool," says Tanya Debose, a fifth-generation Houstonian. "Well over 80 percent of our community is in a flood zone." The flood zone was not naturally occurring; the interchange had created it.

A few days later, an FHWA investigator showed up at Modesti's house. Modesti gave her a tour of her home and walked her around her sliver of the Fifth Ward. "She was like, 'Oh my God, this is crazy,'" Modesti recalled. "I'm like, yeah! What do you think we're trying to tell you?" To Modesti, the investigators seemed overwhelmed by the amount of evidence that they would need to sift through. In January, when she called to ask for an update, they told her that they had gathered much more information than expected. "So it's going to be a longer investigation," they told her. It was starting to feel as if the highway would never get resolved—that she would never be able to relax in her home.

Modesti felt uneasy. The creepy postcards and solicitations had stopped. But nothing had changed; the uncertainty was exhausting. "No one is bugging me. I'm not getting any 'Hey, let me talk to you' messages," she says. "But I'm not okay, because I don't know what's going to happen."

The uncertainty wore on Elda, too. In early February, her house is decorated for Valentine's Day, the kitchen table set with four white placemats embroidered with the word LOVE. Shiny red heart garlands hang on the wall, draped over framed family photos.

Since the letter from TxDOT in 2019, Elda has received no communication. No offer for their house, no notice about when they'd have to move out. The unanswered question—would they have to move?—was a low-grade hum, always audible, the background noise to their lives. She was happy the project was paused because it meant that maybe their home would be spared, but in the meantime it felt as if everything else in her life were paused, too. They wanted to put a new roof on their house. They had finally saved up money to build the second bathroom she had always wanted. "But if they want to throw it away, then what's the point?" she says. "Right now we don't know how much time we have. It's very frustrating." They had worked hard to pay off their home. "It's something that's finally yours. That you can say, this will be for me, for my children later on," Elda says. She chokes back tears. The idea that she and Jesus will be unable to leave something for their children—this is what makes her emotional.

The sky has turned gray by the time Jesus arrives home and joins us at the heart-covered kitchen table. "Have you put any thought into where you might move?" I ask him.

"We're going to have to do something, whether we like it or not," he says. "Here's what I don't understand. They say this highway, supposedly according to the statistics, is the most deadly, where there are the most accidents. And they want to make it bigger. Don't you think the accidents will increase too?"

"They say they are going to fix it," I say, searching for the right Spanish word—*arreglarlo*. "Build a safer highway."

"I don't believe them," Elda says.

Jesus takes me outside to their expansive backyard. Many of their fruit trees were killed in the winter storm a year earlier. Now their branches are brittle and twiggy. There are no leaves to dampen the sound of cars streaming south on I-45.

12

DEMAND

Austin
I-35

One day in June 2014, Jay Blazek Crossley was sitting at his desk in Houston watching C-SPAN. Jay was a public policy nerd, a native of Houston, and a graduate of the University of Texas at Austin's Lyndon B. Johnson School of Public Affairs. He was in his mid-thirties and had just taken over as executive director of Houston Tomorrow, a smart-growth nonprofit his dad founded in the 1990s ("the origin story is that my dad learned about climate change and freaked the fuck out"). That day, the U.S. House of Representatives had convened to debate an annual housing and transportation appropriations bill.

For the third year in a row, John Culberson, a Republican from Houston, had inserted an amendment barring federal money from being spent on a light-rail line that had been proposed in an affluent Houston neighborhood. It was wasteful government spending, he said. His constituents didn't want it. Another Houston representative, the Republican Ted Poe, moved to strike Culberson's amendment from the bill. Houston voters had approved a transit referendum in 2003, Poe said. The people of Houston had made it clear they wanted more bus and rail infrastructure, and it would set a dangerous precedent for the federal government to overrule local voters.

"It's a throwback to Houston when our only transportation plan was to build highways as far as the eye can see and block any attempt to do anything else," Poe said. "We can only build so many roads. We can only build so many concrete monstrosities like the I-10 West corridor."

Culberson sprang to the lectern in rebuttal. "I'm very disappointed and disheartened that my friend Mr. Poe would stand up and offer this amendment and call the Katy Freeway a concrete monstrosity," he said. "The Katy Freeway is my pride and joy. I got it built in five years and three months. It went from eight lanes to twenty-two lanes. That freeway is moving more cars in less time at more savings to taxpayers than any other transportation project in the history of Houston."

Jay scoffed—a highway was Culberson's pride and joy? And then he wondered: *Were* cars moving faster than they had been before the highway was expanded? He looked at Houston's travel time data, comparing how long it took to drive on I-10 during rush hour from downtown Houston to the far-flung suburb of Katy. In 2005, the thirty-mile trip took fifty-two minutes at rush hour. In 2014—only six years after the state had spent $2.8 billion to widen the highway—during rush hour that same trip took seventy minutes, an increase of 33 percent. The highway was wider, but traffic was worse. In May 2015, Jay summarized his findings in a four-hundred-word article that he posted on the website for Houston Tomorrow. The story spread across the state, and then the country, and then the world. Soon, the Katy Freeway expansion had become the most famous example of the phenomenon known as induced demand: If you make it easier for people to drive, more people will drive.

Culberson's amendment remained in the appropriations bill.

Jay eventually moved back to Austin with his wife. He had an idea to take what he was doing at Houston Tomorrow—advocating for sustainable growth—and scale it statewide. "The first problem about the Texas transportation system is that nobody's bothering to ask for change," he says. He started a nonprofit called Farm&City and began showing up at the Texas Transportation Commission's monthly meetings. Back then, he was usually the only person in the room who opposed highway expansion. "There were the people that work for the companies who want the highway contracts and people representing cities and counties. And that was it," Jay says. "It was so weird and uncomfortable. The sense you got when you were there was that these powerful people thought you didn't matter. They were like, why is this guy here?"

At Houston Tomorrow, Jay had fought the construction of the Grand Parkway, a proposed 180-mile loop that would traverse seven counties around Houston. He lost. He watched as the freeway—and the suburban subdivisions that soon followed—paved over the Katy Prairie, which spread along the far west side of Houston at the end of the massive Katy Freeway. "The Katy Prairie was a sponge that held a bunch of water," he says. When Hurricane Harvey hit, that sponge was no longer there. The Katy Freeway expansion and the construction of the Grand Parkway had incentivized real estate development on the Katy Prairie, converting more than sixty thousand acres of grassland to impervious cover. The water had to go somewhere, and so it flooded Houston.

When he moved to Austin, Jay saw the same thing happening. In Houston, sprawl had been a highway problem—the vast landscape opened by limitless roads. In Austin, it was a housing problem. As the price of housing in Austin skyrocketed, low- and middle-income people had left the city in droves, seeking cheaper housing in the suburbs strung along Interstate 35. In 2020, most of the city was still zoned for single-family housing, which meant that it was illegal to build anything other than a detached home on a single lot. Developers couldn't build duplexes or fourplexes without going through costly and contentious rezoning fights, so even as people flocked to Austin, drawn by new jobs in tech and the city's well-branded weirdness, the housing supply remained stagnant.

Meanwhile, former small towns like Kyle, twenty-five miles south of Austin, had exploded in population. Originally founded as a townsite for the International–Great Northern Railroad, Kyle spent most of its history as a sleepy small town, organized around a Main Street called Center Street. In 2000, just over five thousand people lived in Kyle. But when Austin boomed, Kyle did, too. Priced out of the city, people began pouring south along I-35. When Angel and Michael Leverett started looking to buy a house in 2015, Kyle's population was approaching fifty thousand people, most of whom commuted back to Austin on the highway.

Angel and Michael started dating when they were college students at Texas State University in San Marcos, just south of Kyle. They moved to Austin after graduation and rented a two-bedroom apartment in South Austin, a few miles from where Michael grew up. Over the years, their rent crept up, as rent does. But in 2015, when it jumped from $1,080 to $1,560, Angel thought, no way. Several friends owned homes in Buda, a suburb just south of Austin, and they were paying less for their mortgage and property taxes combined. "You're telling me I can drive fifteen more minutes and I can actually own a home and build equity?" Angel said. "That's incredible." They got married and moved in with Michael's parents to save up money for a down payment. But by the time they were ready to buy, houses in Buda were out of their price range. So, like many people before them, they decided they could drive just a little bit farther to find a home they could afford.

It took a while for Angel and Michael to find a house. They submitted eight offers and got outbid on every one. Finally, in October 2016, they moved into a three-bedroom, redbrick house on the northern end of a new subdivision in southeast Kyle. A year after they moved in, Angel decided to leave her job as a sales representative for Nestlé and got hired at the Settlement Home for Children, a nonprofit that supports children leaving traumatic situations. She loved the work. But her new office was in North Austin, more than thirty miles up I-35. Some days, she could get to work within an hour. But sometimes, the one-way commute required nearly two hours of driving. Angel was a cheerful person by nature, but after months of spending almost four hours a day alone in her car, she became depressed. Rage bottled up in her during her drive. She arrived at work angry. She drove home aggressively. At one point, Michael asked her, gingerly, "Have you noticed you're extra irritated lately?"

She hadn't, actually. She started to put the pieces together: Maybe her commute was making her miserable. There was no time to work out, no time to go to happy hour with her friends, no time to make dinner with her husband. "I live for the weekends. During the week, I just spend so much time in my car," she says. After getting furloughed from her job soon after the pandemic hit, she found a new position managing communications for Austin Habitat for Humanity, an affordable housing provider that operated out of a sprawling office in South Austin. The new job cut her commute by more than half. "Thirty to forty minutes one way? It's like, uh, that's nothing," she says. She felt as if she could finally breathe.

As Angel worked with families trying to find homes—line cooks earning minimum wage who drove home to San Marcos, public school teachers who lived in Round Rock—she learned about Austin's housing crisis.

Mostly, this crisis was created by the city's outdated and exclusionary zoning code, which hadn't been updated since 1984, when the city was less than half its current population. In 2012, the city had adopted a comprehensive plan called Imagine Austin that committed to building a "compact and connected city" that could "accommodate more people in a considered and sustainable fashion." The following year, the Austin City Council hired a California-based urban planning firm to update the city's zoning code, a process they dubbed CodeNEXT. By the time I moved to Austin in 2017, signs opposing the effort had sprouted from lawns across the city, reading CODENEXT WRECKS AUSTIN in bright red type next to a cartoon wrecking ball. A year later, when Austin's mayor, Steve Adler, suggested that the city council end the CodeNEXT process—which had "gone horribly wrong," poisoned by "misinformation, hyperbole, fearmongering, and divisive rhetoric"—the city had spent more than

$8 million and six years on the effort with little to show but rancor and a lot of angry yard signs.

In 2019, the city started fresh. The city council asked for a land development code that would create the capacity for roughly 400,000 units of housing, mostly accomplished through upzoning parcels of land, or allowing for greater density (a duplex instead of a single-family house; a thirty-unit apartment building instead of a twenty-unit one). Increasing density would not only make it cheaper to live in Austin—study after study showed that a greater supply of housing would decrease prices—it would also allow people to live closer to their jobs and schools, thus reducing how much everyone had to drive. The city estimated that a new land development code would reduce the daily number of miles driven by Austin households from fifty miles per day to thirty-nine. The new code would reduce the annual cost of transportation for most households from $11,200 to $8,500 and decrease transportation-related greenhouse gas emissions per household by 25 percent.

But Austin did not get a new land development code. In early March 2020, as the threat of a novel coronavirus loomed, a group of nineteen homeowners sued the city, saying it had denied them the right to protest zoning changes that affected their properties. The City of Austin maintained those rights didn't apply in a comprehensive zoning update, which affected every property owner in the city. A district judge sided with the homeowners. Two years later, after the city appealed the decision, an appellate court in Houston ruled that the city had indeed violated state law. Now the zoning code rewrite would have to pass with a three-fourths majority rather than a two-thirds majority. And the council didn't have the votes.

Austin's failed land development code revision and TxDOT's expansion of I-35 were two versions of the same story. "Zoning in most American cities requires car-dependent urban form," Jay says. When there is no affordable housing in central neighborhoods, near jobs and schools and grocery stores, you have to build bigger highways to move people to the places they can afford to live. Induced demand was also a housing story. Removing I-35 required reckoning with this pattern of development. The highway—and the speed it promised—was the reason that people who couldn't afford to live in Austin's city limits could live in Kyle, twenty-five miles to the south, and still consider themselves part of Austin.

"In my heart of hearts, I would love to live in Austin," Angel says. "But it's not realistic at this point." She doesn't want to live paycheck to paycheck, dedicating half her family's income to housing. Better just to drive. Angel and Michael's lives are tethered to Austin by the highway, and that tether is fraying. Michael loves Kyle. He likes the quiet mornings, the quaint buildings on Center Street. And he wants TxDOT to add lanes to I-35. "If Austin continues to grow like it does, traffic is going to be really bad," Michael says. "It's already really, really bad. It's bad because there's not very many options to get from A to B." Their son, Leighton, was born on Christmas Eve in 2020. Sometimes Michael and Angel joke: When he's eighteen years old and they want to take him into Austin, how many hours will it take then?

On a bright Thursday morning in January 2022, Jay walks up the steps of the Dewitt C. Greer State Highway Building. The building is a gorgeous art deco structure, designed by a San Antonio architect named Carleton Adams. After he got the commission, Adams traveled across the country seeking inspiration for his skyscraper, the first in the state to hold public offices. The Empire State Building had just opened in New York City, and Adams modeled the entrance after that famous structure's, with ornately decorated limestone panels and piers topped with eagles framing the stainless steel doors. The building faces the grounds of the Texas State Capitol, and on sunny days you can see the pink dome reflected in its glass windows. Listed on the National Register of Historic Places, the building is unquestionably grand. The architecture—the heft of its limestone— connotes power.

Inside, at 10:01 A.M., J. Bruce Bugg calls to order the monthly meeting of the Texas Transportation Commission. The few people still chitchatting in the back take their seats. There are plenty left open, the room barely half full. Bugg notes for the record that public notice of the meeting containing all items on the agenda had been filed with the secretary of state a week prior in accordance with the state's open meetings laws. According to agenda item nine, the commission would consider rescinding an order from seven years prior that gave the City of San Antonio control of Broadway, an ordinary six-lane street that happened to be designated as a state highway.

When they saw the agenda, city officials in San Antonio were stunned. "Huh?" wrote San Antonio's mayor, Ron Nirenberg, on Twitter. "State is preparing to claim Broadway is part of the TX highway system, 6 years after transferring control to the city. Amid a remodel that was approved by over 70% of SA voters, @TxDOT is trying to put a halt to the project. @GovAbbott—what happened to small government?"

State highways traversed cities across the state, relics of the preinterstate era. As cities expanded and traffic moved to the interstates, those roads were swallowed by urban growth. Boxy apartments sprouted alongside honky-tonk bars and car repair shops and cafés selling green smoothies, until these former highways looked and felt like city streets. By 2013, TxDOT was facing a budget shortfall so enormous it briefly considered converting some rural paved roads to gravel. Instead, to save on maintenance costs, the state decided to turn ownership of these so-called highways over to cities. In 2014, TxDOT said that State Loop 368—known to everyone else as Broadway Avenue—was "no longer needed for a state highway purpose" and that it would begin the process of transferring ownership of a 2.2-mile stretch of highway over to the City of San Antonio.

City officials saw an opportunity. Broadway was a "stroad," the portmanteau of "street" and "road" coined by Charles Marohn, an engineer and the founder of the nonprofit Strong Towns. Streets build community wealth. Streets are where people go to buy things and eat out and grab a coffee and walk around to see what the kids are wearing these days. Roads move people quickly between two distinct destinations; they are corridors designed for speed. Stroads are street/road hybrids, trying to serve two functions but failing at both. A stroad is a high-speed road lined with places people want to stop, making it less efficient for drivers and incredibly dangerous for pedestrians and cyclists. Broadway was a stroad. Anyone who wanted to get somewhere quickly could just get on the highway that ran almost parallel to the north-south road. Why not turn Broadway back into a street? The city came up with a plan to narrow the six-lane road to four lanes and use the extra space to build a protected bike lane and a wide, tree-lined sidewalk. In 2017, the city put the Broadway redevelopment plan to voters as part of a street improvement bond. The bond passed with 78 percent of the vote, allocating $42 million to the Broadway project. Soon, developers started buying up land along Broadway and submitting rezoning requests for new apartment complexes and office buildings. By 2022, more than $760 million had been invested along the corridor.

And then, six years later and seemingly overnight, TxDOT changed its mind. It turned out the state agency had never actually signed the order giving San Antonio ownership of Broadway. It seemed like a minor technicality, as TxDOT's own staff had been working with city officials on the redevelopment plan for years. Why take it back now?

"I would like to present some background information on this agenda item before the commission votes on it," Bugg says. In 2015, Governor Abbott asked the commission "to directly address the top choke points in our major metropolitan areas, giving this commission a clear policy directive to reduce congestion on Texas roadways," Bugg says. This was the only mandate the governor had given the commission. "That's how important congestion relief is to the governor and to the people of the state of Texas . . . I feel strongly that if . . . capacity is reduced from three lanes in each direction to two lanes in each direction, we would allow an action that would be in direct conflict with our clear stated policy to provide congestion relief for the state of Texas, specifically in San Antonio."

Bugg had known for years that the city intended to reduce capacity on Broadway—how could he not? For two decades, he and his wife had owned a two-story redbrick home in Terrell Hills, a tony neighborhood five miles northeast of downtown San Antonio. Their house was four blocks from Broadway. Bugg would have driven on the street countless times as he headed downtown. If he was opposed to the road diet, he could have intervened in 2017, before the city put the issue to voters. He does not explain this sudden concern over capacity, but the most plausible explanation is that Bugg got a phone call from Greg Abbott, who had somehow caught wind of the plan and wanted it quashed.

Many Texas Republicans, including Abbott, believed that car capacity was synonymous with economic development. "Everything we're doing for transportation infrastructure feeds into keeping Texas number one in the nation for economic development," Abbott said at a TxDOT forum a year earlier. Conveniently, some of the governor's top campaign donors were highway contractors. J. Doug Pitcock, Jr., the ninety-four-year-old founder and chief executive of Williams Brothers Construction, Texas's largest highway contractor, had given the governor $4.3 million since 2002. Incidentally, it was Pitcock's company that won the contract for the $2.8 billion expansion of Houston's Katy Freeway. The owner of another construction company, Hunter Industries, had given $1 million to Abbott during that same period.

A few years after his blog post went viral, Jay started thinking about safety. The commission often talked about how much congestion cost the state of Texas. But what about car crashes? he wondered. More than four thousand people died on Texas roads every year. Surely there was a measurable cost to that carnage. Using a methodology developed by the National Safety Council, Jay started calculating the comprehensive cost of car crashes, accounting for the

CITY LIMITS   151

loss of life as well as the lost quality of life for people who suffered incapacitating injuries. In 2021, he calculated that the annual economic cost of crashes in Texas was $29 billion, more than twice as much as the cost of congestion.

When Bugg opens the meeting to public comment, Jay asks the commission to consider the value of human life over travel time. "Between 2012 and 2021, on this 2.2-mile segment of Broadway, twenty-seven people suffered serious injuries and about three thousand people were involved in crashes," he says. "The economic impact of those crashes on this section of roadway is $50 million. The comprehensive cost, when you factor in pain and suffering, was $270 million." What is the congestion benefit that would justify those costs? he asks. The commission does not respond. (Through a spokesperson, Bugg declined to be interviewed for this book.)

Months later, at a policy conference in Austin, a local reporter asked Brian Barth, TxDOT's deputy executive director for program delivery, what exactly had happened in San Antonio. Even as city staff scrambled to salvage their redevelopment plans, to come up with some kind of compromise to add bicycle and pedestrian infrastructure, TxDOT held firm: Broadway must remain six lanes. "Can you tell us what happened with that project and explain TxDOT's stance?" the reporter asked.

"We definitely understand the need, for a lot of reasons, but safety being one of them, to increase pedestrian and bicycle facilities, make sure those facilities are safe," Barth responded. "Just not at the expense of vehicular traffic."

It is a damning statement, revealing the length to which TxDOT would go to protect car infrastructure—over Texans' dead bodies.

13

HOUSING

Houston
I-45

In 2018, when Jasmine Gaston moved with her two children into the public housing complex tucked on a sliver of land between I-69 and Buffalo Bayou, Clayton Homes had seen better days. Nearly 40 percent of the 296 units flooded during Hurricane Harvey, and residents still complained of mold and pest infestations. Built in 1952, apartments didn't have central air-conditioning or heat. But the location was incredible. Jasmine would go for runs downtown, marveling at the skyscrapers glowing against the dusky sky. Her fourteen-year-old son walked to school and rode his bike everywhere else—through downtown, under the highway into the Fifth Ward, and back along the Buffalo Bayou trail.

Two years later, in 2020, Jasmine started hearing from a company called DRA, the same company that had contacted Modesti Cooper. By then, the Houston Housing Authority had agreed to sell Clayton Homes to TxDOT for $90 million and committed to using that money to build two mixed-income communities within a two-mile radius. Displaced residents of Clayton Homes and Kelly Village, another public housing complex just across the highway that would be partially demolished, would have the first right to return.

But the Houston Housing Authority hadn't started construction on any new units. Where were people supposed to go in the meantime?

"DRA's job was to sell the idea of the highway expansion to the residents of Fifth Ward," Jasmine says. "They sold the highway expansion by saying we'd go from public housing to the housing choice voucher program." Housing choice vouchers, known as Section 8 vouchers, are rent subsidies given to people who earn below a certain income, much like food stamps. But unlike nutrition assistance, housing assistance is not an entitlement program. As a result, a housing voucher was incredibly hard to get, even if you qualified for one—thirty thousand people were on the waiting list in Houston. But people were confused. They didn't know when they would get their vouchers, how much rent the vouchers would cover, or where they would go when they got them. Jasmine says TxDOT—and the company it had hired to relocate people—took advantage of that confusion. After Stop TxDOT I-45 knocked on doors in Clayton Homes, trying to organize residents, DRA told Jasmine that if she spoke out against the expansion, she might lose her voucher. "They said to me, if the other side wins, you don't get your voucher," Jasmine says. Other tenants at Clayton Homes and Kelly Village refused to speak out against the expansion for fear of losing access to a voucher. (DRA did not respond to a request for comment.)

In 2021, when Jasmine finally did get her voucher, she struggled to find a place that would accept it. In Texas, it is legal for landlords to decline to rent to voucher holders, who are overwhelmingly Hispanic and Black, as Jasmine is. As a result, many families struggle to find adequate housing and end up moving to poor, predominantly minority areas. "There are a lot of hurdles to go through in order to use a voucher. Public housing is strictly based off your income. To transition people from public housing to Section 8, it's not the same. It's not a smooth transition at all," Jasmine says.

After months of searching, in October 2021, Jasmine moved to the Summit at Bennington, an apartment complex in northeast Houston. Her new apartment was nice. It had central air and heat, a dishwasher, clean carpets. But she hated the neighborhood. "It's a very industrial food desert. There's nothing within walking distance, absolutely nothing, not even a corner store," she told me. Her kids still went to school in the Fifth Ward, but now it was too far for them to walk or bike. So she drove twenty minutes to drop them off and pick them up every day. After she moved, she got a new job as a medical scribe. She worked from home, calling in virtually to take notes for physicians during patient visits. "The relocation did put a limitation on the type of work I could accept," she says. Without any other form of transportation, she had to plan her days around shuttling her kids around. Jasmine started to notice how the highway expansion was self-perpetuating. Displaced from the center of the city, now she had to drive to access almost everything she needed. "The more that they are expanding the city in the manner that they are going about it, you're going to need a car more and more so."

In May 2020, as Jasmine was waiting for her voucher, a white woman in her thirties named Rebecca Winebar went to look at an apartment in a 375-unit complex called the Lofts at the Ballpark, six blocks southwest of Clayton Homes along the I-69 frontage road. Rebecca and her boyfriend, Peter, had been living together in a tiny apartment in a neighborhood called Montrose, but after three months of quarantine they were desperate for more space. When they visited the Lofts at the Ballpark, which spanned three buildings across two blocks, the leasing agent told them that the complex would likely be affected by the I-45 expansion but that there was no timeline for when construction would begin. City leaders were still wrangling with TxDOT, and Rebecca assumed they had a few years before anything happened. The two-bedroom unit was spacious, with a walk-in closet and stained concrete floors—nearly fifteen hundred square feet for $1,750 a month. "That's the best deal in the city,"

Rebecca says. And they loved the location, adjacent to a light-rail stop; Peter didn't have a car, so he could ride the light-rail to his job downtown.

The neighborhood just east of downtown had once been Houston's Chinatown, but in the early 1990s many immigrant families relocated to the suburbs in southwest Houston, and industrial warehouses moved in. In 1999, the city created the East Downtown Management District, a tax increment finance zone to encourage real estate development. Gradually, it came. In 2009, a new stadium for the Houston Dynamo opened, and people began renovating old warehouses into lofts and offices. Restaurants and breweries opened along St. Emanuel Street, townhomes filling in the blocks behind it. The Lofts at the Ballpark, built in 2001 and renovated in 2015, advertised itself as luxury urban living in a "hip Downtown neighborhood" with "convenient access to freeways and public transportation." Minute Maid Park, where the Astros played, was two blocks west. Walk one more block and you'd be at the convention center and then, just beyond that, at Discovery Green. "This trendy community is exactly where you want to be!" the website promised, alongside photos of a sparkling blue pool surrounded by palm trees.

They moved in. Rebecca worked from home, doing customer service for an e-commerce company. Peter worked for a production company and moonlighted as a bartender. In the evenings, the couple went for long walks and explored their new neighborhood. Rebecca often walked a few blocks down St. Emanuel to True Anomaly, a brewery founded by four NASA employees, to hang out on the patio and listen to live music.

On March 1, 2021—a week before FHWA paused the I-45 expansion—Rebecca got a letter in the mail. "The Texas Department of Transportation has determined that the land containing the Lofts at the Ballpark is required as part of an upcoming public right of way project," it read. "Accordingly, TxDOT has begun the process of acquiring the property, which will necessitate the relocation of all residents of the property." (TxDOT had by then already acquired the property.) Even after the federal government halted the highway expansion, requesting that "no further actions be taken on this project that might impact our Title VI investigation and any proposed remedies should the agency find that a violation has occurred," TxDOT hustled tenants out of the Lofts at the Ballpark. Rebecca was able to get an extension on their move-out date; by the time they left in November, there were maybe forty people still living there, she says. It was spooky—the parking garage nearly vacant, the hallways quiet. Three full-time maintenance employees had been reduced to one. She often saw him running around the complex, trying to keep up. Information was hard to come by. "We'd ask, 'What are our next steps?' They had no idea."

It turned out that Rebecca and Peter were eligible for relocation assistance from TxDOT. For forty-two months, TxDOT would pay the difference in rent between what they had been paying at the Lofts at the Ballpark and whatever comparable property they moved to. HDR, the relocation company, sent them a list of possible "replacement" apartments they could move to. They could use the supplemental housing assistance to help pay the difference in rent or take it as a lump sum of nearly $25,000 to use as a down payment for a house. The money was a windfall—the down payment they didn't have. Houston's housing market had gone haywire during the pandemic, so they deferred their payment for a year and moved into another corporate-owned apartment complex a few miles away while they looked for a home to buy.

Six months after Rebecca and Peter moved out, on a Monday morning in early June 2022, a young engineer named Michael Moritz was riding the light-rail to his job downtown, listening to music and staring out the window. On Texas Avenue, just past the Houston Dynamo stadium, he noticed construction fencing around the Lofts at the Ballpark. A six-foot-tall chain-link fence, curtained with opaque green netting, surrounded the complex. On the fencing hung a giant sign: ALAMO DEMOLITION. He sat up straight in his seat. "That's when it got real," he says.

Michael knew that the Lofts at the Ballpark was in the footprint of the North Houston Highway Improvement Project and that TxDOT had bought the building months before the federal government intervened. But the highway expansion had been paused for more than a year—why was TxDOT proceeding with demolition? The light-rail pulled away from the station. When Michael got to work, he called the demolition company and asked to speak to the on-site project manager. "Hey, I live near this building, I'm curious what's happening to it?" he said. The man on the phone replied, "Yeah, we're about to tear down all three of these buildings."

A native of San Antonio, Michael had moved to Houston for a job at an engineering consulting company that did vehicle crash reconstruction. Companies facing litigation would hire Michael and his boss to go out and re-create a crash scene, often long after the fact. Wearing a hard hat and a reflective safety vest, Michael would stand on the shoulder of an active highway and point a 3-D scanner into traffic, re-creating the world in which the accident had occurred. Once, he was sent out to analyze a head-on collision that had occurred on the main lanes of Interstate 45, north of Beltway 8. Just beyond several long skid marks—the residue of tire rubber as the car braked and swerved—there was a massive black char mark from where one of the cars had caught fire and burned. Michael was rattled, the scar seared into his memory. The work was chaotic, loud, and hot. And it was dangerous, standing unprotected on the edge of a highway streaming with cars. "It became abundantly clear to me working in the car crash world that cars and the way that we move was just inherently unsafe," he says.

What Michael really wanted to do was work in renewable energy. So one day, he emailed Michael Skelly, a Houston businessman who had founded some of the largest renewable energy production and transmission companies in the country. Skelly was active in local politics and had started a grassroots group, the Make I-45 Better Coalition, back in 2017. Skelly invited Moritz to his office. When Moritz told Skelly about his work analyzing car crashes, Skelly told Michael: Get involved in the I-45 fight and I'll help find you a job. "I said, deal," Michael says. He started showing up at community meetings and TxDOT outreach sessions. At some point in mid-2020, he was on a Zoom call talking about transportation safety and got a private message from Molly Cook. "How have we not met?" she wrote. Soon, Michael had joined Stop TxDOT I-45.

After he saw the demolition fencing and confirmed that TxDOT intended to tear down all three buildings, Michael texted Molly: We need to have an emergency meeting. A few days later, a handful of Stop TxDOT I-45 members crowded into a tiny, plant-filled living room. TxDOT had purchased the Lofts at the Ballpark in January 2021 in an early acquisition deal, before the project was paused. Legally, it could do what it wanted with the buildings. "So we were like, what do we do about this?" Michael says. It felt like a watershed moment: TxDOT was tearing down hundreds of units of high-quality housing near transit when the project could still be canceled. They decided to organize a protest on the day demolition was scheduled to begin. "But we weren't sure what the message was going to be for the protest," Michael says.

As he talked with Molly and Neal Ehardt, Michael had an idea. "Hey, how many units did they account for here?" he asked. The demolition company had confirmed it was tearing down all three buildings, even though only the front building was in the footprint of the highway expansion. Neal pulled up TxDOT's final environmental impact statement. Buried in an appendix, hundreds of pages deep, there was a table listing the sixteen multifamily housing facilities that would be demolished, including the total number of units displaced. On the last row of the table, the Lofts at the Ballpark was listed at 165 units. But now, it seemed, TxDOT was tearing down all 375, 210 more than it had been authorized to remove.

The black asphalt radiates heat at 6:30 P.M. as people gather in an empty parking lot on St. Emanuel holding white poster board signs stenciled with tidy black lettering: HOUSING NOT HIGHWAYS. CASAS NO CARRETERAS. SAY NO TO A WIDER I-45. Others pass out cold water bottles from Coleman coolers. Molly and her sister, Chloe, unfold an enormous canvas drop cloth, folded in half and spanning twenty feet, so long it requires seven people to hold it up. One side of the canvas is painted with giant red letters: DO NOT PAVE OVER US!!

Four local news crews show up. Michael sets up a tripod, and the journalists hook up camera mics and frame their shots. "Tearing down dense housing next to transit for a highway is the opposite direction we should be going as a city," Michael says. "These are valuable apartments. They were touted as the new urbanism when they were originally built in the early 2000s and now TxDOT has decided they're not worth anything."

The crowd, more than a hundred people carrying signs, begins to march down St. Emanuel to the Lofts at the Ballpark, eight blocks north. Molly wears a neon mesh vest over her red Stop TxDOT I-45 T-shirt and helps direct the crowd along the street. She holds up a megaphone, leading the group in a call-and-response. "Your racist project is sure to fail—my neighborhood is not for sale!" they cry. "When the air we breathe is under attack—what do we do? Stand up, fight back!" The giant canvas sign snakes along the sidewalk. It's a game day for the Houston Astros, so the sidewalks are full of people in orange and navy jerseys, walking to the stadium or congregating at bars. On the patio of a Little Woodrow's, people hold pints and stare at the spectacle on the street.

I weave through the crowd and find Rebecca Winebar, walking along with the protesters. She'd been back to the Lofts at the Ballpark recently to pick up a package that had been sent there by mistake, she says. "The only person there was a security guard for the demolition company. He let me in to look for my package, and the leasing office was abandoned with all the computers and furniture still there. It was super creepy. I wanted to be like, can I take something?" Instead, she just left; her package wasn't even there. It was sad to see the buildings abandoned, she says. "It was March when they were supposed to stop with the project, and it was March when they started telling us all that we had to get out. Why were they so aggressive?" Rebecca asks, as we walk up St. Emanuel. "Why did they have to push it so hard?" It seemed like such a waste— all the time and energy and money spent relocating tenants who could have just stayed put. "All of the legwork of making this happen when this shouldn't be happening at all."

The crowd surges right, crossing St. Emanuel at Capitol Street, which has been closed to traffic because of the demolition. The protesters congregate in the street, standing in front of the construction fencing that surrounds the Lofts at the Ballpark. "Listen, TxDOT, hear our cry," Molly yells into the megaphone. The crowd replies, "When we're bullied, we multiply." Everyone cheers as Michael takes the megaphone from Molly. "I have some really good news that I want to share with you guys," he says. "Right when this event started, the FHWA released a statement asking TxDOT for more information on the demolition at the Lofts at the Ballpark, including the decision to destroy three buildings, when only one was in the footprint. How about that?" Michael yells. The crowd cheers. "We did that. This group right here did that." They clap and whoop.

The loss of housing was a significant impact under the National Environmental Policy Act, FHWA wrote. Accordingly, the agency had asked TxDOT to clarify "key events in the acquisition of the Lofts at the Ballpark, and the decision to acquire and demolish all three buildings, in order to evaluate whether the full impact of the loss of housing from all three buildings of the Lofts at the Ballpark was properly considered and mitigated in the NEPA process."

Walking back to the parking lot where the protest began, in the fading summer light, we pass by True Anomaly, the brewery Rebecca used to walk to. Michael Duckworth, one of the brewery's owners, had heard about the highway expansion when he signed a lease and started renovating the historic warehouse, but he decided the risk was worth it. In 2016, the expansion of the elevated highway that ran in front of the brewery was a vague possibility, not an imminent threat. He and his partners poured concrete, installed new windows, and built a barrel room, sinking nearly half a million into the structure before opening as True Anomaly Brewing Company in 2019. The name comes from an orbital mechanics term— "the part of an equation that you would use to find out your relative position to an object in space," Duckworth says. He talks like a space engineer—"we were able to re-vector pretty quick," he says of COVID-19. They sold growlers during the first uncertain months and hosted live music on the patio as the year wore on. And then, in the summer of 2020, TxDOT published its final environmental impact statement, showing that it intended to consume the entire block of St. Emanuel to widen I-69, which included True Anomaly. Since then, Duckworth had been scouring warehouse space in the area, looking for a place to relocate the brewery. "One of the bigger things that we're losing is the presence, the proximity, the location," he says. "We'll find a place, but it won't be this. That's irreplaceable. It's disheartening to see, not quite a decade into this new growth and revitalization, that they are going to go and lay waste to it."

On the patio, the elevated highway is visible through globe lights strung between poles. Faint electronic dance music wafts from a rooftop club a block away. People on green scooters whiz through the intersection. Couples walk, arms looped loosely around waists, flashing summer skin. So many other cities are desperately trying to bring this scene back to their deserted city centers: the casual chemistry of a late Tuesday night in midsummer. And yet, in Houston, leaders are prepared to pave it over.

Early the next morning, Letitia Plummer, an at-large city council

member, announces that the city wouldn't grant TxDOT demolition permits until the federal government concluded its investigation
into the agency's acquisition of the property. "Things move so
quickly, and we already have a housing shortage . . . and we can't
displace any more people in the city," she says. TxDOT had insisted
that delaying demolition permits "presents significant public health
and safety concerns and would require resources to keep the buildings secured. Vacant buildings such as these have attracted illicit and
illegal activity which would be burdensome to the city, local communities, businesses in the vicinity and potentially exposes TxDOT
to an array of liabilities." A local reporter looked into it: The only
illicit activity reported since January was a single parking violation.

At a city council meeting later that day, Houston's mayor, Sylvester Turner, lays into TxDOT. "You can't take more than what's
needed," he says. "At a time when affordability is a critical question,
the housing stock is in short supply, TxDOT is taking more housing
units than they need. Even if the green light was given on this particular project, I just don't know what TxDOT was thinking. It
doesn't do anything to generate goodwill." The following evening,
Turner writes that TxDOT has informed him that the agency
wouldn't be seeking further permits from the City of Houston.
State and federal properties were exempt from local building codes,
so TxDOT didn't need a permit to tear down the buildings. "Assuming no objection by FHWA, and no demolition permit being re-
quested by TXDOT," Turner writes, "the city has no legal basis to
stop the demolition."

Two months later, on a steamy morning in August, a charter bus
pulls up in front of the Dewitt C. Greer State Highway Building.
The doors creak open and more than fifty people filter out, most of
them in red Stop TxDOT I-45 T-shirts, carrying rolled-up posters
and to-go coffee mugs. Molly directs the group down into a base-
ment meeting room, where they pile tote bags and backpacks, and

then back upstairs to congregate on the steps in front of the building. Over the next hour, their numbers grow. Adam Greenfield and Jay Blazek Crossley arrive. There are activists from Dallas and Fort Worth and San Antonio. Three people from El Paso arrive to stand on the steps, having driven nearly six hundred miles across the state the previous day.

In early 2021, Jay, Adam, and Heyden Black Walker sent an email to twenty freeway fighters in Texas, including Susan and Molly in Houston and Patrick Kennedy in Dallas. "As you know, advocates and grassroots organizations across Texas are pushing back against TxDOT's many highway expansion plans around the state," they wrote. "This frequently requires under-resourced groups to leverage colossal amounts of time and energy without guarantee of victory. But if such groups united and supported one another, it could make the chances of success for any one campaign more likely." Activists from across the state started meeting monthly. On one of those Zoom calls, an idea for a coordinated statewide protest emerged.

Molly stands on a retaining wall and screams into the megaphone. "Safe streets are our right, this is why we have to fight!" The group repeats her chant, bobbing their signs up and down. People wearing business suits and blazers begin to arrive. They walk up the steps and into the building, studiously ignoring the hubbub. As it approaches 10:00 A.M., the group of protesters filters inside. They line up along the eastern wall of the meeting room and unfold the giant white banner they'd painted two months earlier: DON'T PAVE OVER US!! Molly's dad, Mark Cook, holds one corner of the banner, which spans the length of the wall, the vivid red lettering causing a ripple in this hushed, staid space.

Today, the commission will vote on its 2023 Unified Transportation Program, or UTP, a ten-year plan that would allocate an unprecedented $85 billion to widen and maintain highways across the state. Eighty-two people have signed up to give a three-minute comment before TxDOT can vote to approve this plan—less than one person for every billion dollars that TxDOT will spend on highways. "I ap-

preciate all the attention and outpouring of support for our UTP discussion today," Bugg says. But there was an "unusually large number
of individuals wishing to comment." Accordingly, Bugg would limit
the time allocated to each person to one minute. He calls TxDOT's
general counsel, Jeff Graham, to the lectern to remind the assembled
audience how to lawfully give public comment. "This is an extraordi-
nary turnout," Graham says. It is, he repeats, "atypically large." He
reminds commenters that "the decorum needs to be maintained, anyone who causes a disruption needs to be asked to leave."

Back in 1970, a civil rights leader in Dallas had organized a similar
trip to the Dewitt C. Greer State Highway Building. People living in
the Spence neighborhood in South Dallas—named for the five-block
span of Spence Street—got wind that the Texas Highway Department intended to run Interstate 45 through the middle of their
neighborhood, without even an on-ramp for people to access the
highway. "We learned that the damn highway was going to be built
on top of South Dallas, with no way for blacks in South Dallas to get
on the highway," recalled Peter Johnson, a prominent civil rights activist. "Not to mention the fact that it was going to split the Spence
community up." People were angry. "These were just hardworking,
everyday people living in little shotgun, wood-framed houses. But
they were *homes* to these people," Johnson said. Black physicians in
South Dallas rented vans so that people could travel to Austin to
testify in front of the transportation commission, to plead with the
state to spare their homes. "How'd that go?" a reporter asked Johnson in 2014. "We were received by the Texas Rangers. . . . It was a
very hostile situation," Johnson said. "We had to train people about
non-violence before we took them into a situation like that, so we
didn't get somebody bloodied or killed. We didn't take people with
us who weren't committed to non-violence. Because it wasn't a picnic. It was confrontational."

There are no Texas Rangers here today. The protesters do not fear
for their safety. And yet, once again, Black families have come from far
away to plead with these white power brokers to spare their homes.

Jasmine comes to the lectern. She wears a red Stop TxDOT I-45 T-shirt and a gray baseball hat and reads from a handwritten statement, staring down at the piece of paper as she shifts her weight from foot to foot. "My name is Jasmine Gaston and I represent myself in opposition of the I-45 expansion and having it removed from UPT," she says, transposing the acronym in her urgency to get through her statement. "Ever since being pushed out of my former community, Clayton Homes, due to the expansion, commuting my children to and from school has been more time consuming and food has become more scarce. Expanding the freeway only benefits a select few at the top," she says. "But for the rest of us, regardless of race or socioeconomic status, this expansion will lessen your quality of life and cost you more money."

Over two hours, speakers are called to the lectern, asked to state their name for the record, and given sixty seconds to speak. Widening a highway doesn't fix traffic, says Adam Greenfield—never has, never will. Widening a highway will make flooding worse. Widening a highway will destroy housing during a housing crisis. Widening a highway will increase greenhouse gas emissions during a climate crisis. Texans want trains, the speakers say. They want frequent buses and safe bike lanes and streets lined with sidewalks.

Throughout the meeting, the four commissioners sit, silent. Bugg calls each speaker to the lectern, carefully turning over a yellow comment card before announcing who will be next. "Molly Cook, please come forward," he says. "Chloe Cook will be next."

"Cook special, two for two," Molly jokes when she gets to the microphone. No one laughs. For what feels like the hundredth time, she urges the commission to remove funding for the I-45 expansion until the highway can be designed with real community input. "The people crawling up your back today are doing what they have to do to claw their way into a democratic process that belongs to them," Molly says. She sounds tired, or maybe just tired of giving this same speech, only to be met with unflinching silence. But Chloe, Molly's younger sister, is not tired. "Good afternoon, Commissioners. It's me again," she says. Does this commission realize that receiving six-teen hundred comments on a decade-long, $85 billion plan does not represent real engagement? she asks. "We spent $8,000 getting peo-ple here today to talk to you. So stop and listen to what these people are saying. Your hearts have been hardened. Your ears have been closed, and people are suffering as a direct result of the decisions that you make. And it's crazy to me that none of you are willing to take responsibility for it."

Of the eighty-two people who have traveled to testify, all but a few oppose dedicating $85 billion to highway expansion. But the speak-ers do not have a vote. More than two hours after public testimony began and less than thirty seconds after it concludes—before the last speaker has made it back to their seat—Bugg calls the item to a vote. The UTP passes unanimously.

Before the meeting began, a guy in his late twenties named Hexel Colorado had spoken on the steps of the Dewitt C. Greer State High-way Building. He is new to the freeway fight. He grew up in Dallas, worked as a software developer, and hadn't thought much about how he got around until his car broke down and a repair shop quoted him $3,000 to get it going again. "I thought, *Why am I feeding into this mad-ness?*" he said. "I didn't think I would, like, get active about this," he says, gesturing around him—he didn't think transportation advocacy would become his whole personality. Mostly, he just didn't want to have to pay for a car. "But then I realized the effect of car dependence and car infrastructure on my own life," he says. For three years, he'd lived on the edge of a highway and never visited any of the local res-taurants or businesses that surrounded him. Instead, he would hop in his car and seek out some cool place elsewhere. "What I realized was that by choosing where I lived next to a highway, my time, attention, and dollars was being pulled somewhere else that was not present in the place I was," he said. Living without a car had forced him to be present. It had forced him to value proximity. "Highways are the anti-place," he says. "Where highways expand and where highways grow, place is literally destroyed."

14

ACCESS

Dallas
I-345

If Dallas was a laboratory of forgetfulness, the fact most people wanted to forget was that a president was once killed while driving through downtown. In the years following John F. Kennedy's assassination—as the city became known nationally as the "City of Hate"—local leaders leaned into the future. "Dallas's major challenge is to create and shape the future instead of being run over by it," said Mayor Erik Jonsson in 1967. Jonsson, the co-founder of Texas Instruments, set out to reinvent the city. He started with city hall. The Chinese architect I. M. Pei was chosen to design a new structure for the city government, and he conjured a brutalist concrete monolith that leaned over the street. When it opened in 1978, the angled building was a spectacle, writes Mark Lamster, the architecture critic at *The Dallas Morning News*—"a beige aircraft carrier of a building with its angled prow emerging from a vast sea of urban nothingness."

Late in the afternoon on a sunny Wednesday in October, Ceason Clemens walks into this angled prow and takes the elevator to the sixth floor, to the city council chambers. A few weeks earlier, Ceason had been promoted to district engineer, becoming the first woman to lead TxDOT's Dallas district. She's come to city hall to brief the members of the Dallas City Council on TxDOT's plan for I-345—the option selected from the five alternatives presented to the public in June 2021, more than a year earlier. She sits at a table flanked by Gus Khankarli, the director of the city's transportation department, and Michael Morris, the transportation director of the North Central Texas Council of Governments, the region's metropolitan planning organization.

Ceason begins her presentation, flipping to a slide with a map of the highway. "We really went in open-minded and really wanted to hear what your constituents wanted us to study," she says. "The key takeaway from June 2021 as we showed multiple alternatives was 65 percent of the folks that responded to the survey or provided input really wanted to see a highway remain along 345." Despite insisting that public feedback did not constitute an election, when the poll went its way, TxDOT leaned into it. The promised nineteen thousand hours of delay were sufficiently ominous to sway the public's preference.

Since June 2021, TxDOT had conducted an origin-destination study, splitting the city into seven zones and tracking where people began their trips and where they ended them. How would each of TxDOT's proposed alternatives affect travel times for those aggregate drivers? "With the depressed, the elevated, and hybrid, really the travel times are about the same," she says. "We're not going to impact your travel time. With the removal, or making Interstate 345 a boulevard, there is going to be a significant increase in travel times. In the morning, that increase is in between 30 to 40 percent. In the afternoon commutes, that is between 40 to 50 percent."

It was an unacceptable delay. So TxDOT had selected what it called a "hybrid" alternative, which consisted of tearing down the elevated structure, lowering the highway into a trench, and connecting the existing street grid over it. It was a clever bit of branding, implying that TxDOT had somehow found a compromise between a highway and a highway removal. Despite the name, the hybrid option was actually a highway *expansion*. A simulated flyover video showed some sections of the highway increasing from eight lanes to ten. This highway expansion would free up land, Ceason reports—nine acres, in addition to another nine that could be built as caps over the freeway. Some people had said that the boulevard option would "open up hundreds of acres of right-of-way," she says. "That is not the case." It would be more like twenty-five acres, she says—comparable to the eighteen acres enabled by the hybrid option.

"That's not really an apples-to-apples comparison," says council member Paul Ridley, when it is his turn to ask questions of Ceason. Ridley's district spanned downtown and spilled across I-345 into Deep Ellum. A lawyer who had practiced construction law for three decades, Ridley was the city council's resident nerd, digging into the most minute details. Patrick Kennedy had never claimed that the highway itself consumed hundreds of acres. It was all the land that surrounded it, depressed by its presence—the dirt lots, the squares of surface parking, the wasteland of asphalt. And land constructed over a ten-lane highway at a cost of millions of dollars per acre was not "surplus" land; it was land conjured for anyone who could afford it. Ceason reminds Ridley that the City of Dallas would have to buy any surplus right-of-way at fair market value, even with the boulevard option.

"Wouldn't it be fair to say that the purchase cost for existing land would be considerably less than the cost of building structures and decks over a freeway?" Ridley says. Ceason hedges: It depends on what the city wanted to build on that land.

Klyde Warren Park was undeniably a success. A park over a highway was better than no park over a highway. But increasingly, TxDOT was using the promise of deck parks to sell highway expansions. In Austin, the transportation agency had agreed to rebuild I-35 so that, structurally, it *could* be capped. But TxDOT wouldn't pay for it. Instead, the City of Austin would have to come up with hundreds of millions of dollars to cover the freeway with so-called caps and stitches. Now, it appeared, TxDOT was making the same pitch to Dallas. Let us widen this highway, and we'll let you pay to put something on top of it.

In 2016, TxDOT's CityMAP study calculated that removing the highway would cause negligible delays, increasing congestion on nearby thoroughfares by just one minute. What was the source of the enormous increase in traffic delay with these new projections? Ridley asks.

"With CityMAP, we looked at a regional level," Ceason says. "With this feasibility study, we looked at the impact on the users specifically to Interstate 345."

Michael Morris jumps in. Morris had been in charge of transportation for the North Central Texas Council of Governments for more than thirty years, earning him the title "King of Dallas Sprawl" from *D Magazine*. A few weeks before, when I asked him about the travel delay presented at TxDOT's public meeting in 2021, he'd told me, "You can't just go to one meeting" and expect to understand the issue. Now he adopts the same patronizing tone with the city council. CityMAP was "at a policy level, probably from 300,000 feet," he says. The newest models represent "brain surgery," a much more detailed analysis of the I-345 corridor.

"But the CityMAP calculation included a much broader geographic area, is what I'm hearing you say, than your current statistics justifying the hybrid option for 345," Ridley says.

"That is correct," Ceason responds.

"Why wouldn't we consider that broader geographic scope and its effect on travel times rather than just narrowly focused on travel on I-345?" Ridley asks. The highway did not exist in isolation—it was one highway in a city full of highways. People drove on it simply because it was the quickest way to get where they were going. Remove the highway and people wouldn't mindlessly fill up the boulevard. They would find different routes.

"You're not comparing some number at a very aggregate level of eight years ago to brain surgery that's occurring in the work that TxDOT has done for the community now," Morris says.

Why couldn't TxDOT do that "brain surgery" for a broader geography? Ridley asks.

"If that's a request, we can go back and find those people to do that work and see if they're able to do so," Morris says, as if transportation planners were not fungible. Almost an hour later, he offers the same thinly veiled threat that J. Bruce Bugg gave to Houston the year before: Use it or lose it. "You have a small window with the federal infrastructure bill," he says. State transportation revenues were being allocated. "This is why we are working so hard to expedite projects to slot them in this window. So if there's a huge delay, this project ends up not getting consensus, we're moving on to other projects."

Buried within this threat is yet another threat: The city had very little power over what happened to I-345. It was TxDOT's facility, and so TxDOT got to decide. And TxDOT was governed by the Texas Transportation Commission, which had made it crystal clear that car capacity was its first and highest priority when it stopped San Antonio from turning a state highway into a boulevard. And Broadway wasn't even a real highway. It was hard to imagine the commission would allow Dallas to remove an elevated interstate. Without a change in leadership at the state level—without a new governor—the most the council could do was refuse to support TxDOT's hybrid plan.

A week later, Ceason heads back downtown, this time to the J. Erik Jonsson Central Library. The library opened in 1982, part of Jonsson's grand plan to rebuild and reimagine downtown Dallas. As the building ascends, its eight floors gradually step away from Young Street, an architectural response to the tilted city hall across the street.

Tonight, Paul Ridley has convened a town hall to discuss TxDOT's plans for I-345. "It's an opportunity for you to not only learn but also ask questions about your particular concerns and issues," he tells the assembled audience, a few dozen people in suits and work wear. Before the program begins, he turns the stage over to a wildlife services officer who spends an unusual ten minutes talking about the city's coyote management plan. Finally, Ceason steps up to the lectern and delivers essentially the same presentation she gave to the city council a week earlier. The boulevard was eliminated from consideration because of its effect on travel times. Removal would create "a disparate impact on folks in the southern sector who are trying to get north to jobs," Ceason says. Michael Morris had argued in front of the city council that removing the highway would violate Title VI of the Civil Rights Act. "I'm not convinced the boulevard option can be built in the real world," he said. "Under federal rule, you have a disparate impact to the users of the transportation project." Morris offered no legal evidence to back up his claim, and so there it hung, ominous but undisputed.

After Ceason concludes her presentation, Patrick Kennedy stands up to present his rebuttal. "Thank you, TxDOT, for all your hard work on this. I know it's a difficult job and they've been nothing but professional," he says. "I just so happen to believe that highways were never intended to go through cities and would love to take their liability off their hands and turn it into an asset for the city." He glances down at the laptop in front of him. "I'm going to try to set the land speed record for slides per minute," he says. The audience laughs. "I want to give the historical perspective in terms of how we got here, and why I believe we're compelled in the twenty-first century to reduce emissions and revitalize American cities. I also understand that I am suggesting a very radical and different vision and direction than the city has been moving in for essentially the last seventy years."

And so he begins, flipping through slides he has presented dozens of times before. The first slide shows the downtown highway networks of Dallas, Washington, D.C., San Francisco, and New York City, black lines floating on a white background. "Except, those were the highways that were planned in those cities—they weren't actually all built," Patrick says. "In D.C., there was a brutal fight. They were able to protest and prevent the highways from being built in the center of D.C. in the '60s and '70s." The same happened in San Francisco and New York City. Next slide: Black lines disappear from every map except Dallas. Then, in that vacant space, red dots emerge, each one representing a storefront. "Storefronts respond to population," Patrick says. "They also represent vitality." Storefronts show where people spend time, where they earn money, and where they spend money. New York City and D.C. are saturated in red. Dallas remains a network of black lines. The highways went in, and the jobs and businesses went out.

Patrick clicks forward to a slide showing Dallas's downtown highway network, this time overlaid on the city's redlining map. "People say removing I-345 would be discriminatory. That makes me scratch my head." Without exception, every neighborhood that was once shaded red or yellow has a highway cutting through it—including I-345, which had destroyed a Black business district. Since then, city leaders had concentrated poverty in southern Dallas and shuttled prosperity north. Maintaining highways only maintained this segregation. "We should be doing everything in our power to bring jobs and housing closer together to reduce the demand on all our roads," Patrick says. TxDOT had different motivations than Dallas did. "They are trying to move things. We want households. We want economic development, affordable housing."

Ridley returns to the lectern, and Patrick and Ceason sit side by side to answer written questions submitted by people sitting in the half-full auditorium. "How does keeping I-345 align with the city's environmental goals of reducing single occupancy car trips?" Ridley reads. He replies, "It doesn't." Nine of every ten cars on the road are occupied by a single person, Patrick reports—one of the highest rates in the country. "We actually have to build a different kind of infrastructure for a different kind of city to encourage different kinds of movement," he says.

Ridley reads the next question. "In terms of the pushback against removing I-345, talking about making it easier for people to get through this community," Ridley reads, "why is the conversation not around creating places to bring people *to* our community instead?"

"That doesn't happen overnight," Ceason responds. "What is the city doing to create those jobs while also being able to afford living there? We're supportive of living closer to where you work, but what is that transition plan? What happens to the users on the interstate during that time?"

Patrick agrees. Consideration must be given to get people where they are going today. But a massive construction project would not make that endeavor any easier. "If this is five to ten years of pain in terms of the construction, what do we have afterwards? Do we have more of the same, or do we have an entirely different city?" he says. It was worth thinking in radical terms. "When you've been doing the wrong thing for seventy years, there is not a quick fix."

I meet Patrick at a bus transfer center on Malcolm X Boulevard in South Dallas on a frigid Monday morning. I'd asked him to ride with me from here, across the street from Lincoln High School in South Dallas—a few blocks from Lockridge Wilson's home—to Southern Methodist University, just east of the North Central Expressway. The route, which required a transfer onto a rail line downtown, ran roughly parallel to I-345, the channel connecting South Dallas to North Dallas.

The answer to the question of how to get people where they are going without a highway to take them there is in a bus or on a train. The answer to the question Ceason asked—"What is that transition plan?"—is a bus that runs every ten minutes instead of every twenty. The answer requires a different question—not how to fix traffic, but how to increase access.

Patrick is late because his bus is late, and so I hover under the bus shelter, pacing to keep warm. A man sits on a cold metal bench, his sweatshirt hoodie blocking his face against the wind. He's waiting for the same bus I am—the 1, heading north to downtown. The last bus didn't pull in to the transfer station, he reports, and instead barreled up Malcolm X without so much as a pause. "I guess they thought I was sleeping here," he tells me. "But where I'm supposed to wait?" The man stands up. He is going to go stand and wait across the street; he doesn't want another bus to pass him by. Soon, a bus marked "1 Maple" rumbles up, pulling alongside the sidewalk with a creaky exhale, and the man climbs on board. A few minutes later, Patrick walks up, having just disembarked from a southbound bus. The 1 bus comes only every twenty minutes, and so we hover on the sidewalk of Malcolm X, looking south, unsure if we should wait in the transfer center—where the bus is supposed to stop—or follow the man's lead and wait on the northbound curb.

In 2016, after Patrick had made highway removal a citywide talking point, Dallas Area Rapid Transit, or DART, advanced a plan for a twenty-six-mile rail line called the Cotton Belt, extending from the Dallas–Fort Worth Airport east across seven suburbs. "It was a billion-dollar project for what they were projecting was going to be six thousand riders a day," Patrick says. He called up transit planners around the country. "Would you ever build that?" They all said no; the cost per rider would be astronomical. DART had already built the longest rail system in the country, covering ninety-three miles, without many riders to show for it and a generation's worth of debt.

By then, Patrick had been going to city council meetings for years, trying to get city officials to support tearing down I-345. When a member of the DART board resigned, Patrick said, I'll do it. The night before the city council voted on Patrick's appointment, a council member called Patrick and said, " 'Hey, you're not going to screw me over, are you?' They were worried I was too much of an activist," Patrick said. "That I was just going to raise hell."

He didn't want to raise hell. He wanted Dallas to have a functional bus network. For decades, DART had been building rail lines across the sprawling metropolis without any kind of coherent plan, sporadically adding new bus routes. "We'd never done a comprehensive overhaul of the system in the history of DART," Patrick says. When Patrick joined the board, DART didn't have a network, he says; it had a bunch of lines on maps.

Transit functions best when it connects people across densely occupied places. But in Dallas–Fort Worth, space was abundant, luxuriously occupied. The metroplex sprawled across more than nine thousand square miles, covering more area than the states of Connecticut, Delaware, and Rhode Island combined. The area had some of the lowest rates of public transit use in the country, with just over 1 percent of workers using transit to commute to work. For decades, DART had been fixated on rail. Developers loved it; fixed rail lines increased adjacent property values. And real estate development bolstered the property tax base, which cities loved. Trains ran faster than buses, mostly because buses confronted the same problem cars did: traffic. Nearly 90 percent of DART trains arrived on time, compared with only 76 percent of its buses. But DART would bankrupt itself before it could put down enough steel to connect riders across the region.

We hover on the side of the street chatting until a blue and yellow bus stops at the light, heading north on Malcolm X. "Should we cross?" I ask Patrick. "I think we'd better," he says. We dart across the street just as the light turns green. The bus slows to a stop in front of us and we climb on board, sliding into two seats near the front.

The year Patrick joined the DART board, he finally caved and bought a car. He'd made it nine years in Dallas without one. He took the bus to work for a while, but then DART canceled his route when it opened a new streetcar line and the streetcar didn't get Patrick where he needed to go. There was one brutally hot summer where he showed up at every meeting drenched in sweat. He and his wife bought a fixer-upper in Bishop Arts, a neighborhood in Oak Cliff, and he needed to haul supplies from Lowe's. "Eventually it all sort of piled up, and I was like, I can't do this anymore."

Patrick lost his fight against the Cotton Belt, which broke ground in 2019. But that same year, he persuaded the DART board to hire a transit consultant to overhaul the bus system. "Let's take all lines off maps," Patrick insisted. Many of the planned rail lines had been promised to communities in the 1980s. Existing bus lines consumed capacity without serving many riders. They would start from scratch, with a clean map, and redraw every line across the region. DART's network had been built as a hub-and-spoke model, with downtown at the center. But downtown Dallas wasn't where most people worked anymore, and it certainly wasn't where most transit riders lived. "We built this radial network, which means if you're not going to downtown, you got to go in, change buses or trains, and go out," Patrick says. The rail was already built, but bus routes could function as a grid crisscrossing the city, connecting north to south without the bottleneck of downtown. On January 24, 2022, DART launched a total overhaul of the bus network. From one day to the next, the system changed completely. There were fewer routes, but they ran more frequently and for longer, with service from at least 5:00 A.M. to midnight.

It was hard to measure whether the overhaul was a success; the pandemic wasn't over, and DART had struggled to recruit and retain bus drivers. But ridership was "inching back up," Patrick says. This particular bus is sparsely populated, a dozen people scattered throughout the vehicle. We approach downtown, and the tangle where I-30 and I-345 meet looms overhead. We thread under the freeway and pass through Deep Ellum, swinging east along Hall Street and then back west on Gaston. Patrick points out the window at a three-story garden-style apartment complex, the exterior decorated in red brick and cerulean-blue siding. "This is the first apartment complex I lived in in Dallas, when I walked under 345 every day," he says. Past Good Latimer, the street becomes Pacific Avenue. Stacked cubes of black glass tower over the south side of the street, part of a massive mixed-use development called the Epic, which stretched south to Elm Street, occupying the land that used to contain Harry Wilonsky's auto parts store. The north side of the street looks the way it did when Patrick started walking to work—a stretch of surface parking lots punctuated by low-slung buildings.

As the bus passes back under I-345, Patrick mutters, "Forty miles an hour." On the highway above us, cars roll along as fast as they could travel on a well-designed city street. For Patrick, the congestion was proof that the highway didn't work, not that it needed to be bigger. The bus stops across the street from the Pearl/Arts District rail station with a loud exhale, and we disembark and walk across Bryan Street. As we climb onto the platform, a robotic woman's voice interjects, "Train approaching. Now arriving, Arts District station. Doors will open to the left." A train screeches to a stop. Every row is occupied except for one, which we take. Across the aisle, a woman plays loud music on her cell phone. Someone has brought a dog on the train, and it sniffs at my heels.

As the train enters a tunnel running under Central Expressway, I ask Patrick, "What could DART do with more money?" What if TxDOT started funding urban transit projects? It is such an unlikely scenario that Patrick cannot conceive of an answer. "I think Congress is much closer than TxDOT ever will be to giving us more money," he says. The surface transportation reauthorization bill introduced by the House of Representatives in early 2021 would have increased the federal share of transit projects from 50 percent to 80 percent, "putting highways and transit on a level playing field," he says. As it was, the federal government required local governments to pay 50 percent of the cost of transit projects but only 20 percent of highway projects, making road projects a much better deal for cash-strapped municipalities.

In Austin, the cost of Project Connect—the light-rail plan approved by voters in 2020—had ballooned by more than 40 percent. In April, a memo sent to the Austin Transit Partnership board relayed that the estimated price of the twenty-eight-mile light-rail system had increased from $5.8 billion to $10.3 billion, driven mostly by inflation. Construction costs had skyrocketed, as had the price of real estate. "I'm not saying it's a perfect storm, but we've got some things to work through," said one of the project's planners at a July meeting. The system would get built, just much more slowly, as revenues from the property tax increase slowly trickled in. Only a month later, TxDOT would approve its $85 billion, ten-year plan, with 96 percent of its funding allocated to highways.

The train emerges into the daylight, surfacing just past Mockingbird Lane. "This is our stop," Patrick says. We stand up and lean left as the train brakes. Not counting the twenty-minute wait for the 1 bus—which very much counts—the eight-mile trip here took twenty-nine minutes. In a car, it would take half that time.

For Patrick, transit was about more than buses and trains. "How do we build a city for upward mobility? Because a lot of people in Dallas, especially wealthy white people, don't realize that we're closer to the Memphises and Detroits of the world when it comes to poverty and childhood poverty," he says. Even as North Texas boomed and prospered, the city of Dallas—the region's core—had been left behind. One of three children in Dallas lived in poverty, one of the highest rates in the country. In 2018, the Urban Institute ranked Dallas dead last in a study of racial and economic inclusion across 274 U.S. cities. "How do we have an equitable city, a place that provides a leg up for everybody? Part of that is the transit system," he says.

In cities without reliable transit, access to a car means access to wealth. In 2019, three researchers looked at the correlation between carlessness and poverty. Over the last half century, they found, households without cars lost income, both in absolute terms and relative to households with cars. "As society becomes more organized around vehicles, people without vehicles risk being left out of society," they wrote. "This exclusion occurs not just because people with cars can cover more ground more quickly than people without them, but because changes made to accommodate automobiles can affirmatively disadvantage other ways of moving around." Every investment had an opportunity cost. A wider highway meant less space and money for transit and biking and walking—disadvantaging the people who relied on those forms of travel. The study's authors compared the socioeconomic status of people with and without cars in New York City and Los Angeles and found that "carlessness imposes a smaller economic penalty in places that are not organized around automobiles." This was the fundamental challenge of Dallas, a city organized around the automobile. If not having a car—or a highway to drive on—disadvantaged the most vulnerable, then so did continuing to build a city in which you had to have a car to access opportunity. "A landscape that demands vehicles is a demanding landscape for the poor, for the simple reason that driving is expensive," they conclude. Nationally, the poorest households spend almost a third of their income on transportation—which is to say, on driving. Patrick didn't think owning a car should be a prerequisite for accessing the basic things people needed to thrive—jobs, education, food. "That's just not right," he says. "That's not a sign of a well-built city."

15

SMALL

Austin
I-35

It's dark when Delvis Morales arrives at Escuelita del Alma. She unlocks the door to her classroom—*salón cobre,* the copper room—and turns on the lights. Shortly after 7:00 A.M., the first student arrives, bundled and blinking. By 8:00 A.M., eight kids are scattered throughout the room, absorbed in puzzles, crouched over Legos, concentrating on picture books. Every few minutes, a grown-up arrives with a barnacle wrapped in a bright puffy coat glued to their hip. Reluctantly, the barnacles unglue themselves and walk to the back of the room to hang their layers and backpacks on coatracks stuffed with tiny jackets. The tall people leave and the little ones stand, momentarily baffled at their new autonomy. Soon, they are absorbed into their new world. The kids are four and five years old, the oldest kids at the school, brimming with run-on sentences and kinetic energy. Each kid has a laminated flash card with their name on the front and a piece of Velcro on the back. When they arrive, the children grab their name tags from the hub near the door and affix them to whatever play space they've claimed, an anchor to each activity.

A few minutes before 10:00 A.M., when all eighteen name tags have been claimed, Ms. Delvis claps her hands—one-two, one-two-three. The kids snap to attention and furiously begin to clean up puzzles and art projects. They rush to affix their laminated names on the poster board by the door and then sit, crisscross applesauce, in a circle on a rectangular rug decorated with a rainbow of colors, each one labeled in English and Spanish. Ms. Delvis settles on the floor and starts singing. The kids join in almost immediately. "La lechuza, la lechuza, hace shhh, hace shhh," the kids sing. "Hágamos silencio, hágamos silencio, por favor, por favor."

"Muy bién," says Ms. Delvis. "¿Qué día es hoy?"

Hands shoot up. "¡Lunes!" cries one. "¿Sábado?" asks a boy. "Domingo," says a girl, confident.

"Hoy es jueves," Ms. Delvis says, and points at the word written out on the wall. It is a Thursday in November. In November, what do we celebrate? "¡Día de gracias!" the kids intone. A girl named Frida with dark hair and serious eyes interrupts the lesson with an urgent report. When she was eating dinner last night, Santa called her dad on the phone, she says. Santa knows where she lives and how old she is. "Thank you for sharing," Ms. Delvis says in Spanish, and moves briskly forward. "This week, we're talking about our community. What is our community? Our community is where we live. Where do we live?" she asks the class.

"Austin!" the kids shout in unison.

"What things do we find in our community?" Ms. Delvis asks.

Kids throw out words: Police. Firefighters. Library. Bus. Gas station.

"Yes, all of this is in our community," Ms. Delvis says. "Now we're going to talk about the ways we get around our community," she says—*los medios de transporte.* In this group of four- and five-year-olds, travel by fire engine is the obvious and best method of transportation. But there are many other kinds of vehicles, Ms. Delvis says, and each one has its purpose. There are trucks—construction, garbage, ice cream—and buses that take children to school and on excursions. There are cars that parents drive and bicycles that kids ride. Ms. Delvis sends the kids to sit at their assigned activity tables. At the first table, the kids conjure modes of transport out of cardboard paper and glue. They frown, focused, as they cut shapes with blunt scissors. Lucy makes a police car while Lorenzo constructs a speedboat. As each kid finishes their art, Ms. Delvis pins it to a corkboard on the wall: an ice cream truck that sells every kind of ice cream, a fishing boat carrying two fishermen, a pickup truck with a bed full of vegetables. Samuel glues a yellow rectangle to a white piece of paper and holds it up to survey his work. "I have to make a school for the kids to go to!" he cries. He puts the paper back on the table and draws a big green blob behind the school bus, scribbling a bright orange sky behind it.

Soon, it is time for recess. The kids grab their tiny outerwear from hooks labeled with laminated photos and return to stand on the color rug, heads bent in concentration as they aim zipper teeth into plastic sliders. They get lost in giant hoods, inside-out sleeves, backward beanies. When they are sufficiently layered, they walk to the playground in a tight single-file line, hands clasped behind their backs. Once released, they sprint in circles, chasing each other around and around and around, burning the energy that has built up over the past two hours. The space is shaded by four oak trees, just beginning to shed their leaves. The highway roars overhead. You can't see the elevated structure, towering over the building's front entrance, but the sound is everywhere. Trucks revving, tires tripping on expansion joints, pistons pushing out exhaust. The kids yell louder; the teachers lean in closer. Six games unfold simultaneously on the playground, rules made up in the moment. The kids huddle in threes and fours, negotiate bylaws, and then take off, kicking up loose gravel.

Behind the front desk, Jaime Cano and Dina Flores sit side by side, looking at computer monitors as they eat lunch. Jaime is working on the school's food inventory before he shops for groceries on Friday. Dina answers emails, glancing at yet another monitor with fourteen live video streams, dispatches from every classroom. They're constantly checking in, tracking the arrival and departure of children. Every day is the same—pickup and drop-off, lunchtime and nap time, recess and reading—but its problems are different. Sick children, late teachers, broken sinks. "You land running, every day," Dina says. On the reception desk just in front of her sits a clipboard holding a Rethink35 petition with the words "Say No to I-35 Expansion." Several dozen signatures fill four creased pages. "We need to submit those signatures," Dina says to Jaime.

In May 2022, Jaime and Dina met with representatives from TxDOT. "Essentially what we were told is, we don't know yet," Dina says. Even though the agency had released maps nearly a year earlier that showed dashed right-of-way lines running directly through the school, TxDOT couldn't commit to purchasing the building—or leaving it be. Dina recalls them saying, "It could be that we are going to expand, and if we do expand it, we don't know when it is that you guys are going to have to leave and go somewhere else." Did they have any idea how hard it was to relocate a child-care center? How much time it required? "The other thing that I was told was, 'Don't worry, because we'll help you find another place and we'll help you with moving expenses.'"

TxDOT offered Dina $25,000 to help move the school—the maximum amount offered to small businesses. She was insulted. "That's nothing," Dina says. She'd spent nearly $400,000 moving Escuelita from Congress Avenue and retrofitting the new space. It had taken her almost two years to find Richard Linklater's building, and that was back in 2006, before Austin had really boomed. Escuelita couldn't go just anywhere. The state required preschools to meet certain standards—every classroom had to have a sink and toilet, for example. Each child required thirty square feet of indoor activity space and eighty square feet of outdoor activity space. Classrooms had to be on the first floor only, and children younger than eighteen months required their own separate indoor and outdoor space. Jaime had started scanning real estate listings without any real enthusiasm. It seemed impossible to find a building that met all those specifications, especially one they could afford.

The pandemic had hit child-care centers especially hard. In 2019, there were 14,000 licensed and registered facilities in Texas. By 2022, there were just over 13,000—850 fewer providers than before the pandemic. In the spring of 2022, a survey of 91 child-care centers across the Austin region found that nearly every location—94 percent of them—had a waiting list. Hundreds of families were waiting for a spot at Escuelita. Without affordable, reliable child care, parents couldn't work. This cost didn't seem to factor into TxDOT's decision matrix.

The uncertainty weighed on Jaime and Dina. "We're unsure of where we're going to locate," Jaime says. "We're unsure of being able to find something that can accommodate at least this many families. We're not sure if we're going to be able to afford whatever new lease rate." They thought about buying a building, but real estate in central Austin felt out of their reach. "Even if we can find someplace with all of those things, is that going to be a place that is located conveniently for our current families?" Jaime says. "Are we going to have to say goodbye to those families because it's too far and try to recruit from a new area?" More than 200 families had built their lives around the school, the daily rhythm of pickup and drop-off. Two hundred family geographies revolved around Escuelita. Those families cohered into a community that had kept the school afloat for two decades. Moving would disrupt that fragile equilibrium. "It would cause quite the hardship," Jaime says.

After the meeting with TxDOT, Dina realized she couldn't retire. She had hoped to be able to pass the school along to Jaime or her daughter, Alma, but the uncertainty clouded any plans. "It's like, well, I can't exactly sell Escuelita, or do that transition, if I'm not going to have an Escuelita to sell. So now I'm going to have to wait until I can move us to somewhere else before I can retire," she says. She was tired. Educating children was her life's passion and purpose, but it was hard, emotionally demanding work, caretaking all these small, fragile lives. "I am having to rethink a lot," she says. "How is Escuelita going to continue and is it going to continue? Prices here in Austin are so exorbitant that I don't know that we'll be able to find another building that's affordable. Because I have to admit, Richard Linklater made it very affordable for us to move here to this building." If Escuelita had to pay more for its space, families would have to pay more for child care. "If I can't find a place and then it gets to the point where they're going to start knocking it down, the only other alternative is going to be to just close the business."

After nap time, the kids sit in a drowsy circle on the color rug, their faces creased by the folds of flannel sheets, hair matted and messy. Ms. Delvis reads a book called *Lola at the Library* and then releases the kids to choose their own activity. A girl named Nina stands in the middle of the room, clutching her laminated name tag, indecisive. Where to go? Finally, she scampers over to join Lorenzo at a bucketful of barnyard animals buried in brown crinkle-cut shredded paper. In another corner, a little boy wriggles into a full Spider-Man costume. He plays with a girl who has donned a taffeta princess dress. Two girls take over a bucketful of dinosaurs, walking them up a wooden ramp before engaging in battle mid-air.

The room is full of tiny things. All the tiny things are tidily contained in clear plastic boxes, each one labeled in Spanish, a taxonomy of childhood. There are wood blocks and beads, crayons and colored pencils. Popsicle sticks, pom-poms, plastic googly eyes. Magnetic letters, plastic Easter eggs, rubber snakes. Tiny plastic dinosaurs and frogs and bears, each animal in its own container. A plastic container labeled with the letter *H* and containing only small plastic objects that start with the letter *H*.

It is an extremely specific world, tailored precisely to the little humans it contains. In this world, lining up is very important, as is sitting down. It matters when your half birthday is, because here, half of a year is an eighth of a life. The victories are small—zipping your coat all by yourself, being first in line to open the door to the playground, finishing your drawing before time is called. The world of the copper room is filled with animals and colors and shapes. Time is abundant, parceled into ten-minute segments filled with activities and snacks and books. Inside this room, a single feather, glittery and pink, has a place that it belongs.

By 4:00 P.M., the grown-ups have begun to arrive. Nina is the first to go. "¡Adiós, Nina!" the class yells in unison, as she walks out the door holding her grandfather's hand. Frida looks at the clock. "My dad will be here soon," she says, excited. Five minutes pass. He hasn't arrived. She is crushed. "When will he be here?" she asks, her eyes filling with tears. "All the moms and dads will be here very soon," Ms. Delvis says. It feels like an eternity to Frida as she struggles to hold back tears, but after another five minutes her dad walks in the door. Her eyes widen. The tears stop. She stands up to show him her day's drawings. He kneels and asks for a hug. She throws herself into him. A few minutes later, Jaime comes by to do a kid count. "Hola, Mr. Jaime," the kids cry. "Hey, Enzo!" he says to a kid wearing shorts and oversized cowboy boots. "Me gusta tus botas. Qué vaquero." Enzo beams.

The grown-ups start arriving with greater frequency. Every arrival is a disruption, a reminder that this small, contained world is nested within a larger one, vaster and more urgent. The parents do not belong here; they come in carrying cold air and the rush of adult time. A mom arrives to pick up her son, who is not done with the activity he just started. "Come on, Sam," she says. "I don't have time, buddy, we have to get across town." The highway that rumbles overhead feeds this rush—the desire to arrive, a scarcity of time instead of its minute-by-minute presence. The grown-up world is contoured by concrete monoliths. It's hard to reconcile these two worlds: a place where there is a container for every small object, a room where everything happens, sleeping and eating and dancing and reading; and a highway where nothing happens except in-between time.

At 5:30 P.M., there are five kids left in the copper room. Ms. Delvis shepherds them back into their puffy coats and starts stacking tiny chairs against the wall. She turns the lights off and they walk out of the room, single file, heading to the front classroom, where the remaining two dozen kids from across the school await their parents' imminent arrival with Jaime. It is chaos, so many big emotions in a very small space.

Matt Rutledge walks in holding the hand of his six-year-old son, Lucas. His two-year-old daughter lights up when she sees him, and Matt swoops down to pick her up. When Matt and his wife bought their house on Robinson Avenue in 2014, just behind Escuelita, Matt wasn't thrilled to move a block away from a major interstate. But they were first-time home buyers, they had a budget, and they loved Cherrywood. They had moved to Austin from Brooklyn and wanted to live someplace where they could walk around and get to know their neighbors. When they started trying to get pregnant, they didn't think about child care or where they might send their future kid to school. When their son was born in 2016, Escuelita had a two-year waiting list. But they got lucky. Because teachers usually parked on Robinson, which ran behind the school, families who lived on the street got to skip the waiting list. For five years, Matt walked Lucas to school every day. When his daughter was born, she joined the walk around the block to Escuelita.

When Matt heard that TxDOT intended to expand I-35 straight through his kids' school, he started to think about moving. He didn't want to live next to a twenty-lane highway—didn't want to suffer through the construction. "They are going to be bringing down the upper decks, digging deep holes. It's going to kick all sorts of stuff up into the environment," he says. "I don't want my kids breathing that stuff." Instead of moving, he started trying to organize his neighbors to fight the expansion. Brandy Savarese lived with her husband a few blocks south of Matt. Their yard backed up to a vacant lot that would be subsumed by frontage road. Brandy was a light sleeper and was often startled awake in the middle of the night by the *clu-clunk clu-*

clunk of cars speeding along the highway. She was horrified by TxDOT's expansion plans. During the fall of 2021, after TxDOT published renderings showing the highway would encroach more than a hundred feet into Cherrywood, Matt and Brandy knocked on doors up and down Robinson Avenue. "It was a dagger in the heart how many people had no clue. Your house backs right up to this!" Matt says. Even many of the people who had heard about the expansion said, "Well, what are you going to do?"

Now, as we stand in an empty classroom at Escuelita, Matt asks Lucas, "What if the school wasn't here. Would that make you happy or sad?"

"Sad," Lucas says.

"What about the parents, would they be sad, too?" Matt asks.

"Yeah," Lucas says. He nervously chews on the collar of his white long-sleeved shirt. Lucas loved going to school at Escuelita. He learned how to make friends, how to go potty, how to play *lotería*. His teachers gave him lots of hugs. He planted a garden, but the radishes were too spicy to eat. He learned how to pronounce all the letters in the alphabet. Now Lucas goes to kindergarten at Lee Elementary. Matt and Lucas drive there. Although it's only half a mile from their house, it's on the other side of I-35. Lucas doesn't mind the highway. "I like how tall it is. It sounds like rainbows farting," he says, delighted with himself.

"What about your sister—what would she do if the school went away?" Matt asks.

"She would just have to stay at home while I would have to go to school," Lucas says.

"Nuh-uh," Matt says. "I have to work."

"Then we'd have to have Sassy over here," he says. "Or Mom."

Matt reminds Lucas that Sassy and Baboo—his grandmother and grandfather—moved to Mississippi. And his mom works, too. "Oh, okay," Lucas replies. He wriggles his arms inside his shirt, bored. This hypothetical problem of the sudden disappearance of his sister's school is not his to solve. Matt wraps his daughter in her tiny puffy jacket and asks his son to carry her backpack. "Why do I always have to wear your backpack?" Lucas asks his little sister. She has no answer. Together, they walk to the lobby. As Lucas pushes open the front door, the sound of the highway rushes inside.

It's 6:00 P.M., the school silent and dark. Jaime shuffles through a stack of paper, already considering tomorrow. Dina and Alma pack up their things to go home. "Dina is teaching me the art of working with what you have in the moment," Jaime says. "She's like, you can make your plans . . ." He trails off, laughing.

"He'll go and design his diagrams, which teacher is going to cover where," Dina says. "And I'm like, I wouldn't advise that you spend that much time on that. It's not going to work out like that tomorrow."

A few days later, Adam Greenfield sits in a corner of his house, tucked below two shelves lined with books. He wears a collared sweater and a scarf wrapped around his neck, looking unusually British. "Maybe I can just share a few updates before we brainstorm," he says. Rethink35 does not usually meet on Sunday evenings, but Adam called this emergency Zoom meeting after a local TV station reported that TxDOT would begin construction on a portion of I-35 extending south from Austin as soon as Tuesday. "I did reach out to our attorney, Charles Irvine, about this, and we are going to have a call tomorrow morning," Adam says.

In June, Rethink35, along with the Texas Public Interest Research Group and Environment Texas, filed a lawsuit against TxDOT, alleging it had violated NEPA by splitting the I-35 expansion into three separate projects with distinct environmental reviews. The I-35 Capital Express project was in fact three projects, extending across twenty-eight miles. The eight-mile-long central segment had gotten the most attention, given how significantly it would change the heart of the city. But in north and south Austin, TxDOT also intended to expand I-35, adding two lanes headed north and four lanes on the stretch of highway that extended south from Ben White Boulevard all the way to the suburb of Buda. Together, those two projects required more than thirty new acres of land—almost as much as the central segment alone.

Under NEPA, any state agency receiving federal funding for a project must document how the project affects the human and natural environment. That documentation is categorized in one of three ways, depending on the project's perceived effects. Actions that "significantly affect the environment" require a comprehensive environmental impact statement, which quantifies those effects, includes specific ways the agency would mitigate them, and asks for significant public feedback. (The final environmental impact statement for the Houston highway expansion exceeded eight thousand pages.) On the other end of the spectrum, relatively minor projects—like repaving an existing road or repairing an interchange—can receive what's called a categorical exclusion, essentially an exemption from NEPA. Everything in between is considered through an environmental assessment, a relatively concise document, typically a few hundred pages. An environmental assessment leads to either a full environmental review or a finding of no significant impact, which allows the agency to proceed with land acquisition and construction. But because NEPA covers a broad array of government actions, the law doesn't define what makes an environmental or social impact "significant"—whether it's acres of land taken or people displaced—and thus what triggers a full environmental review.

For both the north and the south segments of I-35, TxDOT had completed environmental assessments. In late December 2021, just before Christmas, TxDOT had issued a finding of no significant impact for both segments, which allowed the agency to proceed with land acquisition and construction.

Rethink35's lawsuit alleged that TxDOT had illegally segmented the I-35 expansion, completing lesser environmental reviews for the north and south segments when it should have considered the entire project holistically. Under NEPA, highway projects can advance in segments only if those segments begin and end at rational points, exist with "independent utility" from one another, and don't "restrict consideration of alternatives for other reasonably foreseeable transportation." TxDOT's approval of the north and south segments restricted the agency from considering anything but expansion in the central segment, the lawsuit argued: "A highway—especially one through a major urban and downtown area—needs to be cohesive in order for traffic to flow and congestion be avoided." That cohesion could now only be achieved by adding lanes to the central portion of I-35, the lawsuit argued, even as TxDOT ostensibly considered alternatives as it progressed through a full environmental review for that segment.

Since Rethink35 had filed the lawsuit, not much had happened. TxDOT's lawyers were assembling documentation to argue their case, Adam said. But despite the pending litigation, TxDOT was proceeding with construction on the southern segment. "I emailed Charles to ask, could we file an injunction or something like that to stop TxDOT from doing anything while the judge makes a determination in our case?" Adam tells the group now. "And he said, 'The big issue would be the cost of seeking an injunction.' Charles has been clear that until a judge makes a ruling, our lawsuit means nothing. TxDOT can proceed as if nothing has happened. So I don't know if we have a good legal leg to stand on."

But they had to do something. If nothing else, they wanted to make a ruckus, to show that TxDOT was acting in bad faith by starting construction while the lawsuit was pending. The problem was that no one knew when or where construction would begin—or what it would entail. "It's probably going to be something very unsexy, like putting up chain link and, you know, putting zip ties between the different connectors," Adam says. "It won't be a digger, like uprooting an oak tree and squirrels fleeing."

Miriam Schoenfield, a professor of philosophy at the University of Texas at Austin, interjects with a question. During the pandemic, when she was teaching remotely, she'd taken a road trip across the country, visiting national parks and camping on public land. When she returned to Austin, she saw the city with new eyes. The highway suddenly seemed violently oversized. When she heard that TxDOT intended to expand I-35, she was horrified. "What are our options if this gets started?" she asks now. "Can this be undone? Basically, is whatever is going to happen on Tuesday, is that reversible? Or is it like, this is happening and Tuesday is our only opportunity to intervene?"

"My memory of what Charles has said previously is that once a project is under way, judges are usually loath to mess with it," Adam says. "So I think he would say, once the ship has sailed, you can't bring it back into the harbor."

"If they build the south, won't it be harder to stop the north and central?" Miriam asks.

"Well, not according to TxDOT, because they insist these are independent utilities," Adam says, chuckling.

"But according to reality, because you've got a whole bunch of lanes full of cars," Miriam says. "I'm just trying to figure out, like, how urgent is it for us to stop this on Tuesday? Like, is this the moment where we stand in front of the bulldozers?"

Adam pauses. "Yeah, yeah," he says, slowly. "We should be doing that. I mean, if not now?"

But there were less than two days before TxDOT was scheduled to begin construction—not nearly enough time to get a critical mass of people onto the highway, even if they were able to find out where construction would begin. Anything short of a critical mass was dangerous. "It's all about optics," Adam says. "You cannot physically stop a major highway construction project happening with probably any numbers that we could reasonably muster. It's all about putting up a fight," he says—broadcasting to the Austin community that "this is a really hot issue and people are on fire about this and this is a big deal."

On Tuesday morning, nearly fifty people gather under a white canopy tent on an expanse of grass that runs along the Interstate 35 frontage road. Most of the people are men, wearing sleek wool jackets over suits or bulky work bombers and cargo pants; it is bitterly cold, unseasonable for Central Texas. Rows of folding chairs face a stage adorned with a lectern, five chairs, and a banner that reads, CLEARING THE WAY FOR TEXAS DRIVERS. Behind the stage, two yellow backhoe loaders face each other. Stretched between the raised loader buckets is yet another banner: I-35 CAPITAL EXPRESS PROJECT. A tidy line of beige dirt extends on the grass in front of the diggers. Ten shovels stand erect in the dirt, each decorated with a navy and yellow bow.

It turns out, the only bulldozers Adam and Miriam might stand in front of are ornamental—the two diggers strung with a banner behind a white canopy tent on an expanse of grass. The start of construction turns out to be a ceremonial groundbreaking, a formal affair attended by elected officials and TxDOT leadership. Laid out on folding tables next to cardboard coffee carriers, sugar cookies thick with frosting are screen printed with the words "My I-35 Capital Express" and "#EndTheStreakTX," referencing TxDOT's campaign to end the streak of daily deaths on Texas roads since November 7, 2000. Several local news crews arrive and set up cameras. A reporter from the local NPR station plugs in his mic to the sound system.

A few minutes after 11:00 A.M., after the crowd has filtered into seats, the speakers assemble on the stage. J. Bruce Bugg sits wearing black cowboy boots under his suit pants. It is loud, so close to the highway, the thrum of traffic a dull roar that permeates the space.

In the distance, just beyond a stand of sprawling oak trees, a line of people carrying poster boards emerges, striding over the grass. As they approach, two TxDOT employees walk briskly toward them, rattled but professional. They stop the group a couple hundred feet away from the tent. "We're going to ask you to stay in this typical protest area," says Melissa Hurst. She doesn't clarify the limits of the protest area but asks the group not to interrupt the groundbreaking ceremony. "We are being respectful by allowing you to be here," she says.

The group confers and decides to honor TxDOT's arbitrary request. Kelsey Huse, a volunteer with the group, holds up a megaphone and starts shouting into it. "Corporate greed is what we fight—polluting Texas is not right," she yells, pacing back and forth behind the imaginary protest line. People repeat her words, a call-and-response. Meanwhile, Miriam attempts to reason with Hurst. "It's frustrating that they haven't listened to us. You, TxDOT, haven't given us a real opportunity to weigh in on this," she says. Hurst doesn't respond. "More cars, more pollution—more lanes is not the solution," Kelsey continues, screaming into the megaphone.

Back under the canopy, the roar of cars speeding past on the frontage road muffles the cries of the protesters. Tucker Ferguson, Austin's district engineer, approaches the lectern. "We are thrilled to break ground on the I-35 Capital Express South project," he says. "Today the population in the Austin region is about 2.1 million. We expect the population to double by 2045. What does that mean? It means more users, more people, more residents, and more cars on our highways."

As he speaks, the protesters grow antsy. They march up the frontage road sidewalk and reconvene behind the canopy, standing on a public street. They hold up their homemade signs: MORE LANES = MORE TRAFFIC. FREEWAY WITHOUT A FUTURE. THIS IS A CLIMATE CRISIS. Adam has the megaphone now and his voice wafts into the tent as Bugg stands up to speak. Standing outside the tent, next to the roaring highway, Adam can't hear Bugg, but standing at the lectern, Bugg can hear Adam. "I-35 through Austin is an absolutely vital corridor," Bugg says. But this vital corridor is threatened by congestion, he says. "Which is why this project, I-35 Capital Express, that's going to start here on the south end and we're going to take it all the way up to the north end and go through the center of Austin, this is going to be a $7.5 billion project that is absolutely sorely needed."

"Shame on you, TxDOT!" Adam cries.

"Keep in mind that 93 percent of Texans get behind the wheel of a car or truck as their means of transportation," Bugg says.

"Boooooooooo," Adam yells into the megaphone.

"This is going to be—this being I-35 Capital Express through Austin—is going to be the absolute signature project, not only of the Texas Clear Lanes programs, but for Austin," Bugg says. "Improving the lives of the men and women who live in Austin and have to get up every morning and go to work—"

—"Hey-hey, ho-ho," Adam yells, the group echoing his chants—

—"take their kids to school, get to soccer games on time"—

—"these profiteers have got to go!"—

—"they want to be home, back with their family"—

—"When the air we breathe is under attack, what do we do?"—

—"and not spend all of their time mired in congestion," Bugg says—

—"stand up fight back," the protesters yell.

Finally, after half a dozen officials take the stage and offer their congratulations and gratitude to TxDOT, the power brokers gather in a line behind the tidy mound of ceremonial dirt. Each person grabs a shovel and lifts a scoop. They pose, smiles frozen on their faces, the freeze-frame dragging on just a moment too long. Someone yells, "One-two-three." Ten shovels toss ten mounds of dirt onto the freshly cut grass.

When KUT, the local NPR affiliate, covers the groundbreaking, the reporter includes a quotation from Bugg: "This project, I-35 Capital Express, that's going to start here on the south end and we're going to take it all the way up to the north end and go through the center of Austin." Adam emails the story to Rethink35's lawyer, Charles Irvine, writing, "Is this helpful for making the case that TxDOT sees this as one big project? Because it looks like that's exactly what he's saying." Charles replies, "Useful. Thanks." The statement is not a slipup, an accidental admission of guilt. It is hubris—an inability to conceive of a narrative in which a twenty-eight-mile highway expansion through the middle of one of America's fastest-growing cities would be conceived of as anything other than progress.

TxDOT's Austin district posts a photo from the groundbreaking a few hours later on Twitter. The Rethink35 protesters are in the photo, small but visible in the distance, framed perfectly between the digger's lift arm and the line of smiling dignitaries tossing dirt.

PART III

16

REPAIR

Rochester, New York
Inner Loop

Beth Osborne can't sit still. Her days are stacked with meetings, but even in the lulls between meetings she roams the offices of Transportation for America, which occupies a suite in a building four blocks from the White House in Washington, D.C., next door to the offices for the National Chicken Council. Everyone wants Beth's input on something—the lineup of virtual sessions for a conference the nonprofit is hosting on Saturday, the agenda for a luncheon the following week, the design of a forthcoming report on equity in transportation. Beth listens to podcasts at one-and-a-half speed, and this is often what her days feel like, too.

Finally, she settles into an orange plaid armchair in Steve Davis's office. Steve, an Atlanta native and former photojournalist, got a job at a nonprofit called Smart Growth America in 2006. Two years later, the nonprofit launched a campaign focused on the upcoming surface transportation reauthorization bill. It took Congress four years to pass the law, so what was intended to be a short-term campaign morphed into ongoing advocacy, eventually becoming a permanent program called Transportation for America. "Then the question was, what do we do after reauthorization passes? We started doing all this work at the state level," Steve says. By the time Beth took over in 2016, Transportation for America had grown to include more than a dozen full-time staff.

"Should we reach out to James and just let him know where we're at?" Steve asks Beth.

"No, I'll see him at the Complete Streets dinner. So we can talk to him a little bit there," Beth says. A few months earlier, Beth had a conversation with James Hardy, the former deputy mayor of Akron, Ohio, and a program officer at the Robert Wood Johnson Foundation. When the Reconnecting Communities program opened for applications in the summer of 2022, Hardy worried that places like Akron would lose out to bigger, better-resourced cities—like Austin—that had the staff and expertise to quickly submit applications. He asked Beth to help come up with a program to help smaller cities compete. "I was just brainstorming ideas," Beth says. "And then he said, to make myself clear, would you do this?"

"I saw that Akron has a project that they're trying to get a grant for," Steve says. "The Innerbelt. Do you know about this?" He swivels his chair around to face his desktop computer and pulls up Google Earth, zooming into a swath of highway north of Akron's downtown. "Look at all the traffic," he says, sardonic. The road, as captured by Google Earth's satellite, has no cars on it.

"It's like the Inner Loop in Rochester. When I saw it in person, I asked if it was open," Beth says, staring down at her laptop. "Look at this." She reads the headline of an article published in the *Beacon Journal*: "The Failed Akron Innerbelt Drove Decades of Racial Inequity." She holds up her laptop and pivots the screen around to face Steve. A black-and-white aerial photograph taken in 1978 shows the highway's construction. In the foreground are single-family houses, cars in driveways, and trees in backyards. Just beyond, a cleared stretch of dirt sweeps through the frame, the bulldozed land stripped of any former habitation. "It looks like a bomb went off," Steve says.

"I can't believe there was a day where people were like, you know what we should do? Tear down all the businesses and houses around our downtown. That seems smart. Let's do that," Beth says.

Steve turns back to the Google Earth image. "They've got an awesome new baseball stadium downtown," he says, scrolling along the Innerbelt's trajectory. "They should tear down the whole thing."

When Coalition for a New Dallas launched its campaign to remove I-345 in 2013, Steve walked around the Smart Growth America office with his laptop, showing everyone the website. "The vision of it was so compelling," he says. "That's the next big phase of this whole fight. Every big, successful Reconnecting Communities project so far, or the ones that are on the cusp, are mostly unused segments."

Removing a nearly vacant stretch of highway in a city like Rochester or Akron was a relatively easy sell; it wasn't hard to imagine that land put to better use. Even the Embarcadero was a highway that didn't really go anywhere. But in a city like Dallas, where everything was so far-flung, highways still felt essential. "When will some place say, we're going to keep the perimeter, but the thing that goes through the center of town, we're going to get rid of that?" Steve asks.

"Which is how they were supposed to be built in the first place," Beth says.

In 2013, Beth visited Rochester and walked around the Inner Loop with the city's then mayor, Tom Richards. The highway was eerily empty, a car zooming past every five or ten seconds. In the 1950s, Richards told Beth, the engineers came in and said, "We need to widen the roads to speed up the traffic, especially to ensure that in the case of a nuclear attack we can evacuate the city," she recalled. And that's what happened. "They widened the roads, they sped up the traffic, and everyone evacuated my city and they never came back," he told her. He wanted to bring people back to his city—to live and eat and shop. Beth's enthusiasm for the project only grew. "It was a total rethinking of what was possible in this corridor and what the purpose of transportation was. So much of transportation turns into a desire to keep the vehicles moving," she says. "It's as if we've forgotten that they are supposed to arrive at some point."

In 2013, the Department of Transportation awarded the City of Rochester a $17.7 million grant to remove a 0.83-mile section of the Inner Loop that arced along the eastern edge of downtown. "There's almost nothing I'm prouder of in my time in DOT," Beth says. "It was probably my favorite project we ever funded."

Rochester is a city with old bones, one built in the era of analog, when print photography was a lucrative, thriving industry. Eastman Kodak was founded in Rochester in 1892; during its heyday the company employed fifty thousand people, a quarter of the working population. Before he died, George Eastman spread his wealth across Rochester, founding the Eastman School of Music and the Rochester Philharmonic Orchestra and contributing considerable sums to the University of Rochester and the Rochester Institute of Technology. Xerox was founded in Rochester, along with Bausch & Lomb and what would become the newspaper conglomerate Gannett. The winters were brutal, but Rochester enjoyed a stunning geography, gathered around the High Falls of the Genesee River and spread up toward Lake Ontario. The city could claim prominence in both the women's suffrage and the abolitionist movements as the once home to both Susan B. Anthony and Frederick Douglass. By the 1940s, Rochester was a thriving, growing city.

But like most American cities, that growth started to gather at the fringes as white people vacated the city's core and moved to the spread-out suburbs. In 1939, the Home Owners' Loan Corporation labeled the entirety of central Rochester either "hazardous" or "definitely declining." Northeast of downtown, homes in the neighborhood of Marketview Heights were deemed "of a rather unattractive vintage." The neighborhood would continue to decline, the assessor concluded. All the neighborhoods near downtown would continue to decline—unless the city intervened.

The intervention the city settled on was the Inner Loop highway, which would "cut deeply into the city's traffic problems" and revitalize downtown, bringing businesses and residents back to the center of the city. Opposition to such a project was "incomprehensible," said Robert Sweet, an engineer for the department of public works. "The Loop puts a tight belt around the downtown section, revives the section and prevents the spread of decay to regions outside of the Loop."

By 1952, structures began to be demolished as the highway worked its way east across the city. The city's two daily newspapers, the *Democrat and Chronicle* and the *Rochester Times-Union*, published frequent updates on the highway's progress, rendering the forthcoming road with thick black marker over aerial photographs of downtown, the structures set to be demolished marked with black Xs. The highway was among the first downtown loops to be built in the country, reported the *Rochester Times-Union*. The destruction of homes was largely reported as a story of demolition: who was paid how much to tear down what structures. Some brick buildings were taken apart "brick by brick," while wood-framed homes were bulldozed.

In 1961, Eva Woods was living with her teenage son and dog in an apartment on Joslyn Place when she received notice that her building was going to be demolished for the last segment of the Inner Loop. In January, her landlord stopped collecting rent and tenants filtered out. In March, the city turned off utilities. But still she stayed; she didn't have anywhere else to go. She and her son lit candles and layered sweaters against the frigid cold. Whenever state officials dropped by, Woods would insist she would be out any day now. If she wasn't home, they'd find a notice affixed to her front door. "Leave my things alone until I get moved tomorrow or I'll turn the dog loose—I need my things. Can't buy more." The postman was told, "Don't believe I've moved away. I'm still here." Finally, in May, two movers—"with a nice way of handling stubborn people," reported the *Times-Union*—showed up at her front doorstep and offered to move her for free. The movers started piling her furniture and possessions outside the building. Douglas McDonald, the president of the company tasked with demolition for the Inner Loop, said $55 would be a cheap price to pay to get on with the job. "Someone told me she has her heart and soul in the place," he said. "Well, she can leave her heart and soul there as long as she takes herself out." More than a thousand homes and businesses would be demolished to build the Inner Loop. Entire city streets, including Joslyn Place, were completely consumed.

Fifteen years after construction began, on October 20, 1965, New York's governor, Nelson Rockefeller, visited Rochester to celebrate the highway's completion. In a ribbon-cutting ceremony just west of the Scio Street overpass, the governor declared, "Much more was at stake here than the execution of a highway plan." What was at stake was the very vitality of the city and its downtown, he said. Rochester would become "a national model in the revitalization and resurgence of its downtown area."

City leaders couldn't have predicted that Rochester's population would peak at 332,000 in 1950, two years before the Inner Loop broke ground—that its downtown would continue to enervate. Or maybe they should have foreseen it. The roads they built were designed to shuttle people out of the city, after all. Only seven years after the Inner Loop was completed, in August 1972, the *Democrat and Chronicle* columnist Frank Zoretich wrote that the Inner Loop "chokes downtown Rochester like a too-tight necklace." He ranked the roadway among the city's biggest and most expensive eyesores. The 2.3-mile loop was "a medieval moat" separating the downtown business district from the rest of Rochester.

In 1991, a comprehensive plan proposed filling in the Inner Loop to bring life back to the struggling business district. Eight years later, yet another downtown redevelopment plan called for the destruction of the eastern portion of the Inner Loop. The New York State Department of Transportation, which owned the facility, didn't oppose the removal—seven thousand cars a day drove on the Inner Loop, far fewer than it was designed for—but it didn't support it either. The highway worked just fine, and the state didn't have money lying around to throw at perfectly functional highways. Then, in 2009, the Obama administration announced its TIGER grant program, which made federal funding available for local transportation projects that supported economic recovery. "It was like, whoa," says Erik Frisch, a transportation planner with the City of Rochester. "Suddenly there's money for transformative projects like this." Suddenly the state had a mechanism to consider the project differently— not as a transportation investment, but as a community development project.

Two years after the city was awarded a TIGER grant, the dump trucks started rolling in. The Inner Loop sank eighteen feet belowground, spanning more than a hundred feet across. During the summer of 2016, people flocked to a small brewery on the highway frontage road to watch as dump trucks hauled in thousands of tons of dirt, filling in the trench that had been dug fifty years earlier. In 2018, the highway reopened as Union Street, a narrow, two-lane boulevard.

On a humid Friday morning in August 2021, I meet Erik and Anne DaSilva Tella, Rochester's assistant commissioner for neighborhood and business development, outside Ugly Duck Coffee on Pitkin Street, formerly the Inner Loop's northbound frontage road. "This is the Anne and Erik show," Anne says, as we start walking north. "We've taken this show nationwide." After the Inner Loop East removal was completed, planners from across the country reached out to Erik and Anne, asking how they'd done it. They'd talked to planners in Oakland, Tulsa, Detroit—all of which were in various stages of removing inner-city highways.

"Hold on a second," Anne says. "Let's stop here." We pause in the shadow of a five-story apartment building under construction, the back side a wall of green sheathing. To our right are low-slung brick buildings. Just ahead, on the corner of Pitkin and East Avenue, a navy-brick building contains a flower shop, houseplants crowding the wide, low windows. For decades, the plants looked out over the Inner Loop's sunken trench. Now they face future apartments. "Right here was the East Avenue bridge over the Inner Loop," Anne says. We walk west over the former bridge, pausing just before a wide green bike lane. "We'd be standing in the middle of the expressway right now," Erik says. Instead, we stand on a sidewalk, on the edge of Union Street, which is lined with grass and young trees. Every hundred feet or so there are red metal benches, complemented by yellow metal bike racks. A three-story redbrick apartment building extends to the north, planters and bikes filling small porches.

Once the Inner Loop was filled in and Union Street opened to traffic, the city was left with six acres of developable land. It wasn't at all clear that anyone would want to buy that land. "One of the criticisms was, well, you're going to create all this land and it's just going to sit there," Erik says. "No one's going to build on it." But there was some reason to hope: Even as the city of Rochester declined in population, downtown was growing. In 2000, nineteen hundred people lived downtown. By 2013, when the Inner Loop project began, more than six thousand people did.

The city bought the land created by the highway removal back from the state department of transportation and asked developers to submit bids for seven parcels.

Howard Konar, a native of Rochester and the president of Konar Properties, was one of the developers that bid on that land. The vacant parcels presented a distinct challenge, mostly because of their shape—long and narrow, like a road. When Konar was designing the complex that would become VIDA apartments, he says their initial plans made the interior look like a scene from *The Shining*. "You would have seen this long hallway, longer than a football field," he says, nothing but closed doors on either side. So they broke the long structure into smaller buildings, separated by stairwells and common spaces. On the south end of Union Street was the Strong National Museum of Play, one of the world's only museums dedicated to the history and exploration of play. The museum had been planning a major expansion; the Inner Loop removal gave them the land to do it on. As soon as the land went out for bid, Konar contacted the museum. What if they developed a parcel together? They could build a "neighborhood of play"—a museum district anchored by apartments and town houses and restaurants, lining a new city street called Adventure Place.

"No one would have been able to imagine that it would look like this," Anne says, as we walk south on Union Street, past four-story apartment buildings fronted by tidy landscaping. More than $300 million in private investment followed the $25 million fill-in. By the time construction concluded, there would be more than five hundred new housing units along Union Street and Adventure Place. More than half of those units would be rented at affordable rates to families earning less than the median income.

The shape of the Inner Loop still lingers. There are only a few businesses on the east side of Union Street—who would have opened a storefront on a highway frontage road? Where the highway once swung west, the edges of buildings form a diagonal, accommodating the shape of a disappeared highway. A retailer selling flooring inherited a polygon in front of its parking lot—the liberated space too small to grow anything but grass.

But it is also possible to forget the highway ever existed. Hours later, I walk to dinner at a vegan restaurant just east of downtown. As the sun finally begins to descend, the summer evening late and long, I walk up Alexander Street and turn left on East Avenue. I pause at the corner of Union Street. Lights flicker on in apartments across the street. A cyclist glides along the bike lane, the whoosh-whoosh of pedal strokes audible across the quiet street. In the distance, the Rochester Gas and Electric building is silhouetted against a reddening sky, yellow globes of streetlights floating in the foreground. I hover on the edge of the sidewalk, enjoying the sense of continuity between the block behind me and the one I'm facing.

A highway is a hard thing to perceive. The physical structure is indisputable—concrete and beams, struts and supports, the persistent rumble. On a city map, a highway is a contour line, defining the shape of a place—this, here, is the edge of downtown. It is the artery that circulates energy, the skin that communicates an edge. But highways aren't designed to be experienced or absorbed at human scale. On a highway, we *are* our cars, contained in the hum and clack of suspended rubber wheels, speeding so fast that we can't absorb anything except paint on pavement and the giant green signs that announce where we are going. But highways are similarly hard to see outside our cars. The structures defy human scale, overpasses soaring 125 feet aboveground, the height of a twelve-story building. A highway is a wall of noise, so loud it dulls every other sense. You can't ever see a whole highway. You can only pause above or below and consider some part of it. And once a highway is built, it is almost impossible to imagine it gone. For most of us, highways simply *are*— the essential shape of the built environment, never to be unbuilt. But cities have always been layered places, colonized and disputed and reclaimed. It's all just construction.

After Union Street opened, people started asking, what about the rest of the Inner Loop? In 2019, the city began to study tearing down the remainder of the highway, a 1.5-mile stretch that bounded the northern edge of downtown, cutting through Marketview Heights, which remained a low-income Black and Hispanic neighborhood. Now, Erik says, people in Rochester would be surprised if the city didn't tear out the Inner Loop North. "It's more like, 'Of course it needs to go. But what comes next? What are you going to do to protect us from gentrification and displacement? What are you going to build in its place, and how is that going to benefit us?' It's more an equity conversation than it is a transportation conversation."

In June 2021, Senator Chuck Schumer visited Rochester and convened a press conference just north of the Scio Street bridge, which crossed over the Inner Loop North. "We have a huge opportunity," he said. "That money in the American Jobs Plan can transform neighborhoods and provide the money we need." Removing the Inner Loop North would be "significant" for Rochester. By the time the highway was completed in 1965, "hundreds of buildings were destroyed; homes, offices, churches, hotels, public buildings, parks, and factories, all met the wrecking ball," Schumer said, squinting into the sun. "Now, like a phoenix, we can bring everything back to life. All that harm with the wrecking ball that was done before can be undone."

Of course, harm cannot be undone—it can only be repaired. If you remove a highway, what was once there does not magically spring forth from the earth. A thousand homes do not themselves build.

A year later, Suzanne Meyer and Shawn Dunwoody arrive at the Scio Street bridge early on a Sunday morning. June can be brisk in Rochester as summer slowly summits the long hill of winter, and the air is still heavy with the night's chill. The bridge is closed to traffic, and a pickup truck full of rolled sod backs up into the middle of the street. Suzanne and Shawn and half a dozen volunteers start unloading the truck, lugging the rolled-up grass onto an expanse of gray asphalt. As each rectangle of grass unrolls and thuds into place, it exhales a small poof of dirt, a dusting of cinnamon on the asphalt.

Suzanne is a retired corporate consultant, a white woman who wears dry-fit button-down shirts and lives with her husband in Grove Place, an upper-class neighborhood full of historic townhomes tucked between downtown and the Inner Loop North. Shawn is a muralist who wears paint-spattered jeans and running shoes, a Black man who grew up in Marketview Heights, just north of the highway. Shawn and Suzanne met when Shawn ran for the city council years earlier and Suzanne campaigned for him. They worked well together. They'd both watched as the Inner Loop East was filled in and felt dissatisfied with the outcome, the blocky three- and four-story buildings that had emerged in its place. "They took a moat and built a wall," Shawn was fond of saying. It was better without the highway. But they felt that the city had missed an opportunity to engage nearby residents—something it hadn't done when the highway was built sixty years before.

When the city began studying the removal of the northern stretch of the Inner Loop, Suzanne and Shawn formed Hinge Neighbors to bring people from opposite sides of the highway together to build a common vision for what would emerge in its place. The mostly white residents in Grove Place were used to talking to public officials, offering opinions with the confidence that comes from having those opinions accommodated. But many residents in Marketview Heights had never even attended a public meeting. Decades of experience had taught them that it didn't matter what they thought. Many of the elders who lived in the neighborhood were kids or teenagers when the highway was constructed. They watched as bulldozers consumed whole blocks, evicting residents like Eva Woods. They remembered, vividly, how it used to be and what had been taken from them.

On the Scio Street bridge, sixty squares of unrolled grass become one square—a pop-up lawn, covering one side of the bridge. Shawn lowers two Bartlett pear trees into wood planters painted bright blue and rolls the planters over to the corners of the growing parklet. A man drags a hose onto the bridge and sprays a thin mist of water over the tiny park. A small, orange-breasted bird lands in the new grass. The bird hops around the imported ecosystem, exploring the unfamiliar terrain. For a moment, it's quiet. The music drifting out of the speakers stops. There are no cars on the highway below. The bird pokes its head into the grass and locates a worm, somehow furrowed into the two-inch-deep soil.

Two hours later, noise fills the bridge. Hundreds of people stand in groups of three and four. They sit in lawn chairs, perch on picnic tables. A jazz band plays on a stage built in front of the parklet, and notes from a saxophone waft over the highway. The low rumble of food trucks radiates from the center of the bridge. Two kids stand in front of a giant chessboard, stumped. Families roll up on bicycles. Couples come straight from church, dressed in their Sunday best.

The event, called Live on the Loop, was Shawn's idea. He wanted to bring people together in a shared space, to imagine what the highway might one day become. The previous summer, the city held numerous community meetings as it presented six concepts for removing the Inner Loop North. In late 2021, the city had announced its preferred concept—a two-lane street that would reestablish the city grid and free up twenty-two acres of land: fourteen for development and eight for parks, including a wide swath of green space adjacent to the World of Inquiry School, a K-12 school that didn't have so much as a soccer field. The following spring, New York included $100 million for the Inner Loop North removal in its five-year state budget. "This area has divided Rochester too long," said Governor Kathy Hochul when she announced the funding.

Martin Pedraza and David Everett stand in line behind two dozen people waiting to buy an empanada at a food truck. The two men grew up across the street from each other, four blocks north of here on Ontario Street. David's family moved to Marketview Heights when he was eight years old. Back then, the neighborhood was full of fruit trees, he says. "Every fruit tree you could imagine. Pears, apples, apricots, cherries." There were a few Black families, like his, some Hispanic families, like Martin's, and a lot of Italian families. "It was a melting pot," he says. You could tell the days of the week by their smell: "Wednesday you could smell that Italian spaghetti sauce all through the neighborhood." On Fridays, when the Catholics in the neighborhood abstained from meat, the smell of fish wafted through the streets. There were corner stores that sold fresh vegetables, bakeries, and meat markets. There were hardware stores, places to buy clothes, and places to clean the ones you already had. Everybody had a job and nobody had a car.

Martin used to have a paper route through Marketview Heights. He'd wake up at 4:30 A.M. to pick up rolled-up copies of the *Times-Union* and the *Democrat and Chronicle*. He lugged the rolled papers around in a sack, walking up and down the neighborhood streets. "Then my father got me a wagon. I was *bad* then," he says—bad as in good. In the winter, the wheels would freeze and he'd go back to carrying the heavy newspapers. "Then they started tearing out the homes," Martin says. "I lost my paper route. I didn't really see it, the homes getting torn down. They did it so fast. It was like, boom boom."

"It took a good part of the neighborhood away from us," David says.

After high school, David left Rochester and played in the NBA for two years—a year for the Portland Trailblazers, a year for the Phoenix Suns—before returning home to teach, eventually earning a master's in early childhood education. At sixty-nine, David is still lean and athletic. He wears a black tracksuit and high-top sneakers. Martin wears a paint-spattered yellow T-shirt that reads "Lewis Street Center Alumni" and camo-patterned cargo pants. For decades, the Lewis Street Center anchored the neighborhood. Kids went there after school to play sports or study. Grown-ups gathered to vote or organize food drives.

Both men are skeptical about the city's plan to fill in the Inner Loop. They wanted the highway gone, but they didn't see how its removal would benefit their neighborhood. The city had asked for community input numerous times. But Martin didn't feel as if that input ever went anywhere. "It's like, every year they come to us, for the last I don't know how many years. They go, what would you like? We tell them. The following year, the same questions. The same answers." They wanted single-family homes that people could afford to own. They wanted businesses—a grocery store, a hardware shop, a Laundromat. There were more than a hundred vacant lots in Marketview Heights, Martin says. Why not start there?

The city had made all kinds of promises, David says: "We're going to build this up. There's going to be housing." But would it be housing that people who lived there could afford? "We've seen this trick before," he says. Sixty years ago, planners sold a vision of progress, promising that urban highways would rescue abandoned cities and bring new prosperity. Instead, they bulldozed neighborhoods like the one Martin and David grew up in. Now different planners were promising that removing those highways would revitalize the same neighborhoods that had been torn apart for an earlier vision of progress. They are justified in their skepticism.

A year earlier, as we walked along Union Street, I asked Anne and Erik how the city could build single-family homes that people in Marketview Heights could afford to buy. There are lots of ways, Anne said. The city could provide construction subsidies to home builders. It could subsidize the cost of land, sell the surplus right-of-way for $1, and require that developers build affordable homes. Rochester was working on a citywide program to support low-income homeownership. But it was still figuring out how to structure the program—to make sure that a home would remain affordable for future buyers, even if the neighborhood around it started to gentrify. "Could the city turn the new land over to a community land trust?" I asked. A land trust would retain ownership of the land under the home, protecting it from future speculation. It could, she said. Everything was possible. "It all depends on the political will," she says.

After we finally get our empanadas, David and I stroll around the bridge until his knee starts bothering him and he heads home. I find Martin talking to Nancy Maciuska, another Marketview Heights native. A few years earlier, Nancy helped start the Lewis Street Committee, a group of longtime residents who had come together to articulate a vision for their neighborhood. Now the two are arguing about that vision—namely, whether it was being ignored. "Right now, the homes in here is $700 or $800 a month," Martin says. "Affordable. Once they do all this, guess what?"

"The prices are going to go up," Nancy says. "That's why we want to be a part of the planning process." Nancy believes change is possible—that the community will benefit from the Inner Loop removal.

Martin is cynical; he doesn't believe the city will follow through on its promises. "It's going to happen regardless of what we say. I have this thing. Don't piss down my leg and tell me it's raining," he says.

Nancy disagrees. "Our voices are being heard. We're not being pushed out," she says. In early 2021, Hinge Neighbors and the Lewis Street Committee, including Nancy and Martin, presented a plan to the city's then mayor, Lovely Warren, that included specific requests for the Inner Loop North removal. They wanted the city to rebuild the neighborhood with single-family homes on small lots. They wanted a business district, places for people to shop and work. They wanted traffic-calming features on residential streets—stop signs, speed bumps, wider sidewalks—and more trees. "We believe that the Inner Loop North fill-in is prime for reparations," said Miquel Powell, a member of the Lewis Street Committee.

"This presentation is not falling on deaf ears," Warren had responded.

"We are being heard, but we're not," Martin says now. "You gotta watch out."

Nancy turns to me. "See, this community has been hurt so much. When they stripped this neighborhood away from us, they just knocked all these homes down and relocated all these families. So I can see his anger and I can understand. But if we don't stand and if we don't make our voices heard, then we're never going to be heard. I know that we're all skeptical. We're all nervous about what's going to happen, but at least we tried."

"We got to try," Martin says, nodding, softening. "You know, like they say, in order to win the lotto . . ."

"You gotta play it?" Nancy asks.

"You gotta be in it," Martin says.

17

LAND

Austin

I-35

In 1867, the Sanborn Map Company started making detailed maps of American cities, documenting the precise locations of buildings and property lines to help fire insurance companies assess their liabilities. Every decade or so, the company updated its maps, sometimes by pasting tiny pieces of paper over lot lines, drawing new buildings, and updating descriptions in small slanted capital letters, creating a palimpsest of the built environment. When the Interregional Highway was constructed through Austin, it swung west as it crossed the Colorado River, cutting across a residential neighborhood before joining East Avenue. A 1940 Sanborn map of Austin was updated some years later, after the Interregional Highway was built. The highway came through the city as a ribbon of clean white paper etched with sweeping black lines, covering the tightly rendered street grid below. Behind the translucent paper highway, the shape of the former city was still visible, the rectangular property lines and small yellow boxes marked D for dwelling. On Clermont Avenue between East Avenue and Waller Street, more than a dozen rectangular lots were covered, almost a full block. Just past the paper highway, at 1103 Clermont, the mapmaker sketched a rectangular frame dwelling on the front half of the property.

By the time the Guadalupe Neighborhood Development Corporation, an affordable housing provider, bought the property in 1999, the little frame house at 1103 Clermont had seen better days. Built in 1938, the house looked as if it were in the midst of a long exhale. Mark Rogers, GNDC's executive director, went to talk to the family that lived there to tell them about the upcoming sale. An older woman answered the door. "You can stay here if you want," he told her. The nonprofit served low-income families and Mark guessed that she would qualify.

"We do like this little house," she said.

"Is it bad being next to the highway—the noise and so on?" he asked.

"No, you get used to it," she told him. "The only problem we ever had was somebody lost a tire off their car and it bounced right into our yard."

The house was a pain, Mark says—no central air-conditioning, ancient plumbing. And it was a small house on a giant lot. So around 2018, long after the woman had left, Mark started drawing up plans to demolish the vacant house and replace it with two brand-new ones, both of which would be rented to low-income families. He applied for city and federal affordable housing funds, as GNDC did for all its projects. He hired an architect and submitted a site plan to the city. He got building permits. He hired a contractor. "Then we found out we're in the wipeout zone," he says. In the fall of 2021, he learned that TxDOT planned to take 1103 Clermont to expand I-35 as it swept into downtown.

When Mark moved to East Austin after earning a doctorate in art history at the University of Texas at Austin, the city was still segregated by I-35. Nearly sixty years after the city had passed its comprehensive plan that created a six-square-mile "Negro district" in East Austin, most people of color still lived east of the highway. Hispanic families had clustered south of Seventh Street, while Black families lived mostly north of Eleventh. That began to change in the early 1980s. "Austin was going through one of its booms. People downtown looked over and said, wow, that's right across the highway and it's cheap land. Let's turn residential streets into commercial streets," Mark says. A local priest approached neighborhood leaders and said, "You guys are going to lose this neighborhood unless you do something." So the neighborhood organized, eventually creating the Guadalupe Neighborhood Development Corporation. The non-profit started buying homes at foreclosure or at cost, rehabilitating them using city or federal affordable housing funds, and selling those homes back to people from the neighborhood. "There was a fortress mentality—let's have a house on every block," says Mark, who took over the nonprofit in 1994. If a homeowner decided to sell the home, GNDC had the right of first refusal to buy it back and resell it to another low-income buyer. They weren't worried about the home appreciating in the meantime. "No one wanted to live over here anyway," Mark says. "These people did because that's where they'd always been."

In 2005, Mark got a call from a couple who had purchased a home from the nonprofit two years earlier for $105,000. They wanted to sell and had already gotten an offer—for $213,000. It was far more than GNDC could afford. "We realized this doesn't work anymore," Rogers says. His realization was prescient. In 2000, most homes in central East Austin cost $80,000. Two decades later, the median sales price would hit $650,000. "We need something else, something that gives us more control over things." A law professor at the University of Texas at Austin named Heather Way was working with affordable housing nonprofits across the state that were struggling with the same question. "How do we ensure that these homes that the community is investing in remain affordable for the long term?" she says. The answer, she thought, was a community land trust.

The first community land trust in the country was established on a farm in rural Georgia. In the early 1960s, civil rights activists outside Albany were struggling to build a movement of sharecroppers, Black farmers who worked and lived on white-owned land. An activist and reverend named Charles Sherrod had been organizing sit-ins and voter registration drives, knocking on doors across the state. But as he recalled in a 1981 interview, he kept hearing the same thing: "What are you going to do if I get kicked out of my house?" The threat of eviction was real. "The only solution that one could come to would be that we have to own land ourselves," he said. And the only way they could afford to own land was to own it collectively. In 1969, Charles and his wife, Shirley, co-founded New Communities, which organized the purchase of 5,735 acres outside Albany—at the time, the largest parcel of Black-owned land in the country. They planned to build five hundred homes, a school, and a health center. Fifteen families moved into existing structures, and many more showed up to help build the fledgling community. Together, they built a greenhouse, nursery school, and grocery store. The farm thrived for seven years, until a drought hit in 1976. When the community applied for an emergency loan from the local Farmers Home Administration, Shirley remembers, a white supervisor said, "You will get a loan here over my dead body." In 1985, they lost the farm to foreclosure.

By then, community land trusts had spread across the country, becoming an increasingly urban phenomenon. In cities, housing is unaffordable mostly because land is. A 2017 study found that, globally, rising land prices account for 80 percent of the increase in home prices since World War II. This is why gentrification has such inertia. Once the engine of capital revs up and investors begin to speculate on a plot of land's future value, the machine is hard to stop. The community land trust model offered an elegant solution: By separating ownership of a house from the land beneath it, you could ensure the home remained affordable—permanently.

Mark worked with Way and a state senator to define and enable community land trusts in the state's property code, finally passing a law in 2011. In 2012, GNDC built the first community land trust home in Texas: a mint-green bungalow with three big windows and a white wraparound porch on the corner of Willow and Navasota Streets. Today, 115 families rent homes from GNDC and another 29 live in homes they own. GNDC retains ownership of the land, which it leases back to homeowners for $25 a month through a ninety-nine-year ground lease. Another 119 community land trust homeowner-ship units and 152 rental units are in the pipeline, Mark says. Every single one will be occupied by a family with generational ties to East Austin.

Late in the summer of 2022, Mark met with representatives from TxDOT. Half a dozen TxDOT employees and consultants sat around a table. As they talked, Mark scribbled notes on the back of the maps they'd given him. TxDOT would publish its preferred alternative in January 2023. Appraisals would follow soon after. Construction would begin in late 2024 or— Mark's notes stopped there. "Probably was going to say early 2025. I didn't even finish writing."

GNDC planned to build two homes at 1103 Clermont, Mark told the TxDOT representatives. They had building permits, funding from the federal government, a contractor. "I said, you know I don't believe timelines. Even when I give a timeline, I don't believe it. I don't believe your timeline. So we're going to go ahead and build them."

Mark scribbled TxDOT's response: "Not going to tell you to build, not going to tell you not to build."

"Why build the houses if they're just going to get torn down?" I ask him.

"Why not?" he says.

"Because you have other homes to build," I say.

"It took a lot of work to get this to this point," he says—more than four years of planning and wrangling and financing. "Why would I stop now?"

"Is there any part of you that's just like, screw you, TxDOT?" I ask.

"I can't say no to that," he says. "I'm kind of belligerent, I've been told." At least he used to be, he says. He's mellowed out as he's got-ten older. "There's still a little bit of the David against Goliath sort of thing. Just 'cause you're big and mighty and have the resources to do whatever you want to do, doesn't mean that I should just kowtow to that. Cower down and say, okay, well, we're not going to do what we're supposed to do. Our mission is to get people housed in affordable housing."

GNDC had nine hundred people on its waiting list for rental housing, all with generational ties to East Austin—people who had been priced out, families who wanted to come back. Another two hundred people were waiting to buy a community land trust home. Mark's job was to provide housing. Eleven hundred low-income people from East Austin needed affordable housing. If he could provide two families with a place to live, even for a few years, that's what he would do.

At the end of Ninth Street on a hill overlooking I-35 perches a single-story blue home, a gable-roofed Victorian built in 1917. When Mary Helen Lopez was growing up in the 1960s, she used to sit on the porch and watch the highway. She got herself a pair of binoculars and watched ambulances race up the highway, lights flashing. She'd watch as paramedics pulled people from cars on stretchers. Mary Helen didn't remember life before the highway, but her mother, Helen Lopez, did. One afternoon, Helen was in the kitchen cooking when she heard explosions down below; workers were dynamiting the path of the highway, rocks launching into parabolas that clattered on their roof. Mary Helen's father, Sabas, ran the Lopez Drug Store a few blocks away, one of only three Mexican-owned pharmacies in the city.

Without air-conditioning, the family slept with their windows and doors open, a breeze filtering in through the screens. When cousins and aunts spent the night at their house, they'd complain about the noise from the highway. "I became immune to the sound," Mary Helen says. "They would say, how can you sleep with this humming of the road?" For Mary Helen, it was like sleeping next to a river. But it was dusty, she says. "Oh my God, dust, dust, dust, all the time."

The dust coated the front porch, filtered inside through the screens while they slept.

In the late 1980s, a group of investors started buying up property around the Lopez house, eventually assembling eight acres east of I-35 between Seventh and Eleventh Streets. In 1991, Delbert Bennett, the real estate broker who had assembled the tract, persuaded the city council to rezone the land to allow a developer to build a massive shopping center. Mary Helen's mother refused to sell. In 2005, a group of University of Texas students produced a documentary about the Lopez house, which had made local news as the only holdout. "They told me, this house is old. It is," Senora Lopez said in creaky Spanish; by then, she was seventy-eight years old. "I didn't say it wasn't. I've done a lot of repairs here. I've spent lots of money fixing it up. I want to die here."

Eventually, the mall proposal withered, but the city never reversed its zoning change. So in the early years of the twenty-first century, as gentrification finally crested the barrier of the highway, developers started building eleven- and twelve-story apartment buildings around the Lopez house, which started to look like the little house carried away by balloons in the Pixar movie Up.

By then, Mary Helen had bought her own house a few blocks away. After her mother died in 2005, Mary Helen rented the Lopez house out to a cousin. A few years later, an electrical fire gutted the house, and it sat vacant. In 2015, Mary Helen called Mark. She was on the board of GNDC and wanted to know if the nonprofit might be able to help fix up her family home. Her mom had fought so hard to keep it; she wanted it to remain in the community.

After GNDC bought the house and restored the interior, Mark started thinking about the 0.17 acres of land the Lopez house sat on. The land was still zoned for a shopping mall, which meant GNDC could build a residential tower up to seventy feet tall. Mark could be cranky—belligerent some have said—but he was also incredibly creative, unbound by conventional wisdom. Who said you *couldn't* build an apartment tower in someone's backyard? An early concept drawing showed a rectangular five-story apartment building with the Lopez house perched on the building's roof. Unbeknownst to Mark and Mary Helen, the city zoned the Lopez house as historic, hoping to help preserve it. "That meant we couldn't put the house on the roof," Mark says. In 2020, GNDC submitted plans to the city for a project it called La Vista de Lopez: a seven-story tower with twenty-seven units, all of which would be rented to low-income seniors. Directly in front of the tower sat the Lopez House, which would be a community center for the building's residents. "People have said we're crazy, and it's pretty much the truth," Mark says.

What was preventing GNDC from building a home for every person on its waiting list was not money, not really; there was money out there to help nonprofits construct affordable homes. It was land. Vacant lots the same size as the Lopez property were being listed for $1 million. Most people on the GNDC waiting list earned less than $40,000 a year. Even someone as relentlessly creative as Mark couldn't make that math work.

A few months before he met with TxDOT, Mark had been invited by the city to participate in an I-35 stakeholder group. Over the decades, he'd participated in too many I-35 working groups to count. This one was called Our Future 35 and was focused on the so-called caps and stitches that the city intended to build over I-35, the deck parks and cross streets that would cover stretches of TxDOT's depressed highway, reconnecting East and West Austin and "mending the divide created by the original construction of I-35." TxDOT wouldn't pay for these caps, to be clear—that was on the City of Austin. The state transportation department would simply build the highway in such a way to *allow* for the possibility. An early estimate had put the cost of capping roughly fifteen acres at $800 million.

Mark went to a few Our Future 35 meetings, and then he couldn't stand it; he was about to become belligerent. "Everyone is throwing out the words 'healing,' 'equity,' 'justice.' I said, 'So you tell me, who is putting up money to bring the people who have been displaced from this area back?'" Mostly, the people who had suffered the injustice of the highway's original construction—those who had found themselves living on the wrong side of a wall—were gone. They had either died or moved out to Pflugerville or Kyle, where housing was cheaper. "We have a list of almost nine hundred households waiting for rental housing. We have a list of close to two hundred people waiting for ownership opportunities. All of them, I mean all of them, have ties to East Austin. When this plan has that element in it, then we can start talking about healing and justice and equity."

But you have to be an optimist to build affordable housing in an unaffordable city, and so Mark oscillated. He looked at a rendering of a cap between Eleventh and Twelfth Streets and thought, well, maybe. "Could they put housing on that? Yeah, it'd be pretty low density. The more dense you get, the more infrastructure you need to hold it up on top of this. It makes the Lopez project look relatively easy, compared to that," he says. "I'm crazy enough to think it's possible. And if you think that, then it might be possible." But then he swung back the other way. "There's going to be so many forces, each silently working against affordable housing going along I-35," he says. It would be great if rebuilding I-35 freed up land. He could get behind that; he could build on that. "But when it really came down to what land was going to be freed up, it wasn't. In fact, they were going to be taking land from people." TxDOT would be taking GNDC's land. It would demolish two brand-new affordable homes to build a twenty-lane highway. The city would put a park on top and call it justice.

The justice was in the land. "Buy land," Mark Twain once famously wrote. "They aren't making it anymore." Removing I-35—or burying it—did just that. Reconnect Austin had calculated that the city could reclaim thirty acres of land if TxDOT tunneled the highway and built a boulevard over it. GNDC was preparing to build twenty-seven apartments on a fifth of an acre. Imagine what it could do with thirty acres. "Unless they're ready to say they're putting money into getting people who have been displaced back here somehow, providing land, then it's all just talk," Mark says.

18

LEVERAGE

Houston
I-45

O'Nari Guidry tells me to turn left. We're on Gregg Street, where it slips under Interstate 10 and into the Bottom, the southern sliver of the Fifth Ward. As I pause at the stop sign, I glance out the passenger window to see Modesti Cooper's house, the pale blue structure towering over its surroundings. I point out the house to O'Nari. Until I called her, she had no idea that TxDOT was planning to expand the highway that displaced her family sixty years ago. "That's one of the homes they're going to take," I say.

"Oh, I'm surprised," she says. In its newness, its angular modernity, it's as if Modesti's home should be imbued with some power that O'Nari's family house never contained. I turn onto the frontage road, away from Modesti's house, and we pass a row of brand-new white-and-navy townhomes, barely beyond the highway's grasp. I drive another block until the frontage road swerves just slightly to the south and becomes Buck Street.

This is the street O'Nari grew up on. "Stop here," she says. "No, wait, go down a little bit more. All of this used to be homes, homes, homes." Now it is mostly vacant lots, overgrown with long grass and shaded by bending branches. "Okay, here," she says. I pull over on the south side of the street, next to a sagging one-story bungalow, the porch beams canted away from the house as if leaning into a strong gust of wind. She points out the front windshield, across the street, to a narrow grove thick with pines and oaks and crape myrtle. Through the trees, we can see cars streaming along the interstate. "I used to live right here," she says. "Where?" I ask, unable to conjure what she's seeing. "Where the cars are!" she exclaims and starts laughing. "Where the cars are traveling. That used to be our house." She laughs loudly, from her belly. When she was a teenager, this new reality was devastating, but so many years later it no longer upsets her; she's driven this street so many times that it is no longer shocking to see the place she grew up covered in concrete.

"The thing of it is," O'Nari says, as we continue along Buck Street, "they could have not come this way with that freeway." There was a factory a few blocks to the south, she says, surrounded by vacant land. "Factory over Black people," she says. Her voice carries no hint of outrage; this was simply the world she grew up in. What did the factory produce? I ask. "I'm trying to remember," she says. "Curley's mind is so much better than mine. He's the one with all the memories."

In 1967, the year after Interstate 10 opened through the Fifth Ward, O'Nari and Curley got married. They moved into an apartment on Sumpter Street. O'Nari started doing clinical rounds at a nearby hospital and earned a college degree. Curley got a job at Shell Oil in southwest Houston. The early years of their marriage weren't easy. Curley thrived on the energy of the Fifth Ward, the clubs and nightlife. O'Nari wanted a house, someplace quiet to raise their children. Eventually, after they had their first son, the couple bought a ranch-style home in a neighborhood called Scenic Woods in northeast Houston. They had another son. They went to work, and came home, and helped their kids with their homework. They had twelve good years and one bad year, O'Nari says. "And one bad year can wipe out a lifetime of good years."

After that one bad year, they divorced. As soon as her younger son graduated from high school, O'Nari left Houston. She'd been itching to leave ever since she graduated from Wheatley. If not for Curley, she would have moved while her boys were still in school. "I felt like they needed their father. I don't think Curley's been out of Houston more than five times in his entire life. He's a real Fifth Warder," she says. "But that was not my goal, to live and die and only see Houston, Texas." She started her career as a labor and delivery nurse, which she calls her "love and passion," but after she left Curley, she'd become a dialysis nurse. The pay was better and so were the hours. Her shift didn't start until 11:00 A.M., so she had time to get her boys ready for school and prepare dinner. A neighbor would keep an eye on them after school, until O'Nari got home from work at 7:00 P.M. When her boys finished high school, she found work as a traveling nurse. She moved to Atlanta but hated the city, wrapped in just as many highways as Houston. She went to D.C., but couldn't stand the cold. She had made her way to St. Croix, in the U.S. Virgin Islands, when her son was incarcerated. He'd gotten married straight out of high school, just like his parents, and had two children. O'Nari wanted to help take care of them, so, reluctantly, she moved back to Houston.

As she helped raise her grandkids—and then, eventually, her great-grandkids—she traveled. She went to Israel, Australia, Italy. She worked her way through Central America—Costa Rica, Honduras, Belize. "I would have moved to Belize were it not for my kids," she says. She was the only one in her family with the "wild hair" to see the world. "My grandmother wanted to do something other than work in the fields. She used to always tell me, 'Baby, it's a wide, wide world. Don't just stay in one spot.'" Her grandmother couldn't travel the world. But O'Nari could.

"Did you ever get remarried?" I ask.

"Oh no," O'Nari says, laughing. "I'm done with marriage, baby. It's a wrap for me. Once was enough."

A few years back, O'Nari's older son started to wonder about the highway that had taken his mother's house. He'd grown up watching cars stream along the interstate from the living room window of his grandparents' home, the house Curley grew up in. "But I don't think he got the big picture of the Fifth Ward until he got older," O'Nari says. "It went *ding ding ding,* about five years ago. That it was houses there, but now it's this big freeway. He couldn't believe it." He told O'Nari, "It's just like if they came and bulldozed homes in my neighborhood." And she said, "Precisely."

O'Nari's life, as she narrates it, is wound up in the neighborhood around us. This is where the pharmacy was, here the grocery store, there was Simpson's Barbecue. Today, the Fifth Ward consists of mostly older homes interspersed with vacant lots. And then there are the town houses; O'Nari calls these "the gentrification people homes." New, boxy structures tower in stark angles over faded bungalows and sagging row homes. They are fortresses, without porches to look out on the neighborhood and sight lines inside. "When I grew up, it was just people everywhere. This used to be full of people, full, full to the rim." Now the streets are empty, resoundingly quiet.

We continue down Lyons Avenue and swing left onto Jensen Drive, pausing briefly at Jensen and Nance Street. The intersection is a forest of concrete, a grove of giant corrugated beams. The canopy above is a tangle of lanes, some branching east to Interstate 10 and some north to Interstate 69. A few weeks earlier, another Fifth Ward native, the Reverend James Caldwell, had given me replications of two grainy black-and-white images, aerial photographs of this same intersection. The first was taken in the 1950s, before the highways came. Nance Street cuts across the frame west to east, crossing the gentle curve of Jensen Drive. Buffalo Bayou swings north in the bottom left corner of the image. Mostly, what the picture shows are homes, densely packed on city blocks—hundreds of tiny rectangular structures. One of them is the home James had grown up in with his parents and six siblings. His dad worked as a longshoreman on the docks of the Houston Ship Channel. When James was in school at

Bruce Elementary, just down the street, a burly kid named George Foreman used to steal his lunch money. As a teenager, James talked his way into Club Matinee, a famous blues club that hosted musicians like Ray Charles and Aretha Franklin. This first image shows all that, the granularity accessible only through the memories of the people who lived there.

The second image is only freeway. Nance Street and Jensen Drive are still visible, as is the swoosh of Buffalo Bayou, but they are overlaid by a tangle of lanes, a dozen channels of curving concrete. The massive interchange of I-10 and I-69 completely fills the frame. The land below the elevated highway is scraped clean, grass-covered blocks captured in soft gray scale. By the time the second picture was taken, James's parents had moved to northeast Houston. They were renters, and so they were offered no relocation assistance. At the time, James didn't think much about what was happening. He lost friends, neighbors. He watched whole city blocks get wiped out. But he was a teenager. He didn't know the same thing was taking place in Black neighborhoods across the country. Later, though, in his thirties—after he stopped doing and dealing drugs and started preaching—it hit him what had been lost. "You almost feel like you're somewhat invisible. Your history has been taken away from you," James says. "You can't change what already is. You can never replace what the freeways have taken over. What you do is try to prevent something like that happening again. I try not to use that old preacher cliché too often. Faith is an action word. You got to do something," he says.

James desperately wants to find images of the Fifth Ward as he knew it because he wants to communicate this memory, to make visible the world that an increasingly smaller number of people can conjure. He wants to say, it wasn't always this way. Which is to say, it could have been different.

This is also what O'Nari is trying to communicate to me—the place that once was, the life these streets once contained. How full and familiar it all was. It starts to rain as O'Nari and I head up Lock-wood to Lyons Avenue back to the Nickel Sandwich Grill, where we'd met. O'Nari is quiet, the only sound the pitter-patter of fat drops on the car roof. Houston needs rain and we both peer up at the sky. One reason the Fifth Ward is a good place to live is that it doesn't flood, she says. It doesn't flood and it doesn't lose power. After Hurricane Ike, she lost power for nearly three months and had to move back in with Curley. After that, she bought herself a genera-tor.

A few hours later, I go by Curley's house. After he and O'Nari di-vorced, he bought the house next door to the one he'd grown up in on Orange Street. Despite his initial uneasiness about the highway when he was a teenager, over the years it seemed to him like an over-all positive for the neighborhood. It brought convenience, accessibil-ity. "It made not just getting to work more accessible, but it made finding jobs more accessible," he told me the first time we met. "I'd rather see it like it is now than looking back if it had stayed like that. A lot of stuff came to us through the I-10. I couldn't imagine where this traffic would be going."

But now, a few months later, as we stand at his living room win-dow and look out over the cement, wet and wiped clean from the afternoon rainstorm, Curley wants me to know that he has changed his mind about the highway. The new highway expansion wouldn't affect him, but he worried about TxDOT coming into his neighbor-hood and taking people's homes. When the highway was con-structed, it had given the Fifth Ward access to the rest of Houston. But it also took something fundamental away from the neighbor-hood. "We lost a lot of talent. When I say talent, I mean, kids that were growing up here that were talented. The freeway moved a lot of them out. The ones that were able to get enough money to move and go into other neighborhoods in the better parts of the city: They split and it left less people here. You know what I'm saying? Like they dipped into the talent pool and they went elsewhere." When the highway displaced all these promising young people, it took away the neighborhood's hope, he says. "Everybody was affected by the I-10."

The last house on Orange Street—a slate-blue two-story house, three doors down from Curley—floats in the no-man's-land between the end of the narrow residential street, which stops at a metal guardrail, and the I-10 frontage road, which sweeps past in a single, wide lane. On a humid, cold Tuesday in November, Molly Cook arrives with a minivan full of party supplies and realizes the only way to access the house's driveway is to go back up Orange Street, drive all the way around the block, and turn right on the freeway access road, which passes the driveway.

As she unloads the van, dragging a cooler heavy with soda and sparkling water inside, Kendra London unfolds two plastic tables and pushes them back against the living room wall. Kendra grew up in the Fifth Ward, raised by her grandmother Ella Morris—the woman who woke up in the middle of the night to find a house on her street. A few years earlier, Kendra had started a nonprofit called Our Afrikan Family, which worked "to give the unseen and underestimated long-term residents of the Fifth Ward community a voice." As TxDOT barreled forward with its plans to expand I-10 through the Fifth Ward, Kendra started hosting monthly bike rides in the neighborhood. When the weather was nice, two dozen people gathered on the corner of Lockwood and Lyons Avenue, straddling bicycles. Together they would glide through the neighborhood, visiting its landmarks as they threaded back and forth across the highway. She wanted people from across Houston to know about the Fifth Ward, to bring people back to its streets. In October, she signed a lease for this two-story house on Orange Street. It would be a community center for the neighborhood, hosting classes and workshops, providing an afterschool space for kids and a meeting space for organizers.

As dusk fades to night, the highway becomes a hazy blur of red-and-white refracted light. People start arriving, walking down Orange Street where they have parked their cars on the grass-covered shoulder. Michael Moritz rolls up on his e-bike. Susan Graham arrives with her son, Scott, along with his wife and two kids. Within an hour, dozens of people stand shoulder to shoulder downstairs, waiting to fill paper plates with fried chicken and green bean casserole and potato salad. Outside, a group huddles around a steel fire pit, flames jumping within. Jasmine Gaston arrives late, frustrated with herself for being late. Since leaving Clayton Homes, she has stayed involved with Stop TxDOT I-45, frequently traveling to Austin with Molly to testify in front of the Texas Transportation Commission. "Are you tired of fighting?" I ask.

"No," she says, simply. She loves it. It gives her a sense of purpose. "I really think it's important to show up and make your voice heard. Especially now that there's digital media, that stuff lives forever. There's a record. It lasts even after we testify. You can see that we were there saying, 'No. We don't want this.' The more you can speak out, the better it is."

That summer, her son was accepted into a magnet high school in Midtown, an affluent neighborhood southwest of downtown. It was a good school, and he had worked hard to get in. But driving both of her kids to school and back home to work now took Jasmine more than an hour every morning and afternoon. "We care about his education. I'm not sending him to any school," she says. "We're trying to get him into college."

Molly calls everyone inside. She stands on the staircase that runs through the middle of the living room. "I just wanted to say welcome—where's Kendra?" "Behind you," Kendra says, perched in the shadows just above Molly. "Kendra, where are we right now?" Molly asks.

"Our Afrikan Family Community Center," Kendra says. People cheer and clap, the sound reverberating in the packed room. "This is truly a blessing, to have everyone here," she says. "The coolest thing is to see all the different age-groups that are in here together. There's always a gap of information that's missing between the youth and the elders. That's our motto: each one, teach one. So whatever I know, you should know for free. Whatever you know, I should know for free," she says. "If you don't teach me, I'll just eavesdrop and learn." The crowd laughs.

Molly takes over. "If anyone is not well acquainted with Stop TxDOT I-45, we started as an idea from Susan Graham," Molly says, still standing on the stairs. "She went to a community meeting about I-45. She walked out and said, this needs to stop. She got a little GoFundMe together, put some yard signs up, and now we are an entire organization that is recognized by statewide governing bodies, that is known to powerful entities across Texas, including of course TxDOT themselves."

An older Black man named Walt raises his hand. "I have a question. What position are we in as far as stopping this I-45?" he asks.

"Good question," Molly says. "Right now the project is paused by the Federal Highway Administration." The project had been paused for a year and nine months, Molly reports. The investigation into alleged civil rights violations was ongoing. "So we are waiting right now for the Federal Highway Administration—they've done their research, they came to town, they talked to people—and right now they are doing closed-door negotiations with TxDOT," Molly says. "That's the way things have always been done; we've been very vocal about how we don't agree with that. But it's where we're at right now. We're waiting to hear how TxDOT voluntarily changes the project to make it less discriminatory."

In the meantime, homeowners like Modesti Cooper and Elda Reyes remained in limbo—unable to move, unable to settle. Modesti had planned to come to this gathering, which was just across the freeway from her house. But she had fallen asleep at 8:00 P.M. A month earlier, a rideshare driver had turned left onto a one-way street and T-boned her car. She walked away, seemingly fine. But within a few weeks, her back started bothering her, pain shooting up her spine. She went to see a chiropractor. "Weren't you in a car accident?" he asked. Maybe that's why her back was hurting. The pain was constant; it felt as if someone were sticking pins in her back. She lost her appetite, had trouble sleeping. Finally, in December, she saw her general practitioner and learned she had two slipped disks.

"Even sitting here is painful," she says a few weeks later, when we meet for lunch at Toot Suite, a café on Commerce Street between Chartres and St. Emanuel, just east of downtown. The sprawling white-brick building had once been a Ford car factory, then a nightclub. Now it was a trendy café serving French macarons and Cobb salads. This would be the historic building's final act, because it was slated to be demolished for the highway expansion. On the east side of the restaurant, floor-to-ceiling pane-glass windows fill with a view of concrete beams and the underside of the I-69 elevated deck; as in Modesti's house, the presence of the highway is inescapable.

The year was almost over, and Modesti still hadn't heard anything from FHWA. Over time, her worry had worn away into acceptance. If she had to move, she would move. But she wasn't going to worry about something she couldn't control. "It's like a layoff," she says. "You don't want to know that stuff a year from now. Let me live my life until that point." Once, in Iraq, she arrived at a new base only to be told that layoffs were imminent. "They told us whoever has been deployed for the longest was staying. The people who had just gotten to the base had to go somewhere else," she says. She started packing up her stuff. When the commander finally announced the layoffs, her name wasn't called. "Here I am, packing all my things, thinking I'm going to leave," she says. "A year ago, I was told I had to leave this house and guess what? I'm still here."

Despite her injury, Modesti had been busy. She was always busy. She was still cooking and catering, delivering bowls of gumbo across the city, and preparing her holiday menus. She was taking care of her eighty-year-old grandmother, who'd recently gotten sick with COVID-19.

In October, Modesti completed a home construction course

through a Houston company called Builder's Academy. The class toured construction sites around Houston. They learned how to pour a foundation, how to check if the concrete was adulterated. They talked to house framers and heard from people who installed insulation. Shortly after she'd returned from Iraq, Modesti registered an LLC and started scouring real estate listings. She bought older houses and townhomes and renovated them, earning a narrow profit on the flip. She bought vacant lots and submitted building plans to the city. Over the past year, she bought three lots on Bond Street in the Fifth Ward and started to build yet another house from the ground up. She liked working in her neighborhood, and it was easier to visit a construction site when it was nearby. But she also wanted to be accessible to the people who bought her homes, to be known in her community. She wanted to help make it better, block by block.

One of the lots on Bond Street already had a house on it when she bought it. "It was in pretty bad shape," she says. There was frayed electrical wiring, sinking floors, doors barely hanging on hinges. She mapped out a new interior layout and worked as the general contractor on the remodel, driving up Waco Street almost every day for six months to supervise construction. Her teenage nieces stayed at her house during their summer vacation, so as the house neared completion, she put them to work laying mulch in the backyard. When it was done, she sold the house for $179,000. The woman who bought it told Modesti that she'd been searching for a home for months but couldn't find anything she could afford. Modesti could easily have listed the home at a higher price: During the pandemic, the median home price had increased by 40 percent in Houston. But playing the market didn't interest Modesti. Her goal was to build homes that people could own for less than $200,000. "My goal is to help the community and make sure everyone becomes a homeowner," she says. She wasn't trying to get rich. "My goal is to help a family or two a year," she says. She scrolls through her phone, pulling up before and after photos of the house on Bond Street. "This is what it looks like today!" she exclaims proudly. "Day and night. It's real cute. Hell, I probably thought I was going to end up living in it. Because I had nowhere to go."

The Fifth Ward was full of distressed homes—houses that people had inherited from older family members and didn't know what to do with. "People don't want to get rid of them, but they don't know what to do," she says. She loved the idea of bringing these old houses back to life. "I don't want to tear anything down. I would rather keep something there and revitalize it. Keep the neighborhood alive," she says. "I'm trying to save homes and my house at the same time. I'm not trying to be like all these people coming in here and destroying what we do have. If you learn how to develop a neighborhood, then maybe you'll keep these damn highways away. And tell them, hey, there's other things you can do to a neighborhood besides put a highway in it."

After lunch, we walk outside to a darkened sky. To the west, behind the elevated freeway, clouds accumulate, thick with rain. The wind picks up, blowing dust into our faces. "You better get where you're going," Modesti says by way of goodbye.

I take my chances and walk down St. Emanuel toward Texas Avenue. Just before the light-rail stop, I pause, disoriented. I am looking for empty space, but still, it surprises me. The front building of the Lofts at the Ballpark has been demolished, the lot scraped clean. Where there was once a four-story building—where there were porches filled with succulents and patio chairs—there is only sky. A single bulldozer rolls from one end of the lot to the other, pushing amber-colored dirt in front of it.

Stop TxDOT I-45's protest had spared the two back buildings at the Lofts at the Ballpark from demolition. Within two weeks of the protest, on July 7, a right-of-way agent for TxDOT gave a tour of the complex to a staffer in the mayor's office and the director of strategic planning at the nonprofit Coalition for the Homeless. Both agreed "that the two eastern blocks could be salvaged for either homeless or work-force housing," reported Allen Douglas, the executive director of the Downtown Redevelopment Authority, to Eliza Paul, TxDOT's Houston district engineer, in an email the following day. They "need to line-up their teams to make the prudent cost assessments necessary, based partly on what we've seen and on our conversations with you and your team. They know they must work swiftly." In October, TxDOT confirmed publicly that it was "exploring other feasible options for alternative use" of the two vacant back buildings. It was a win for Stop TxDOT I-45. And yet, despite the expansion's uncertain future, 160 units of housing had already been demolished.

On a cold, overcast Sunday in mid-December, a six-member mariachi ensemble plays on the banks of Houston's White Oak Bayou. As runners pass on the hike-and-bike trail, two men dressed in traditional *trajes de charro* strum guitars while two trumpet players press quick notes onto brass valves, the sound sharp and full. People arrive bundled in jackets and hats. They queue up for hot chocolate. Kids run over to cornhole boards. Volunteers with Stop TxDOT I-45—their signature red T-shirts visible under unbuttoned jackets—walk through the growing crowd, handing out stapled-together booklets. On the front cover, a silver ornament reflects a scene of Houston's downtown skyline wrapped in highways, with a meandering channel of water glinting below.

Before highways contoured the city's geography, Houston was defined by its bayous. The creeks that snake southeast across the landscape collect water from nearly two dozen distinct watersheds before eventually emptying into Galveston Bay. As White Oak Bayou approaches downtown Houston, it threads back and forth under I-10 for almost four miles, passing through forests of towering beige beams and weaving under canopies of concrete. In 1966, when Congress created the federal Department of Transportation, it also prohibited the secretary of that new department from approving "any program or project which requires the use of any land from a public park, recreation area, wildlife and waterfowl refuge, or historic site."

Today, Stop TxDOT I-45 has gathered people to rally support for a petition asking the City of Houston to acquire and designate the entire White Oak Bayou Greenway as a city park. Designating this span of White Oak Bayou as a park wouldn't necessarily halt the I-45 expansion, but it would require TxDOT to document how the property could be avoided.

"How much fun is it to be in this space right now?" Molly calls to the crowd, sitting on blankets and in lawn chairs, after the mariachi band stops playing. Ally Smither takes the mic. Ally, who manages the group's communications, also happens to be a professionally trained opera singer who has performed at the Kennedy Center. When her husband, Austin—also a classical musician, also a volunteer for Stop TxDOT I-45—got into a PhD program in New York City the year before, Ally decided to stay behind in Houston. "I'm not leaving until this fight is over," she said. Now Ally asks people to open their caroling books. Tucked between paragraphs about the I-45 expansion and induced demand are lyrics for Christmas carols in English and Spanish.

The crowd shifts closer to the band that has assembled next to Ally; a violinist perches on a metal folding chair, surrounded by three guitarists. "Sleigh bells ring, are you listening?" they begin. People raise their voices, easily united by the familiar melodies. A few of the people singing had never heard of Stop TxDOT I-45 until they strolled up and heard the sound of mariachi music drifting over the trail. Now they too join the chorus. One woman wraps a blanket around her legs, shivering. Another disappears into her hoodie. People are cold but they stay. They sing. Their voices carry across the bayou and through the trees. In the pause between chords, when the singers inhale and the guitarist stops strumming, the sound of the highway rushes back—a low, persistent rumble.

The next morning, Molly wakes up early and opens Twitter on her phone. Late the night before, a local reporter posted that there would be an "important news conference" regarding the North Houston Highway Improvement Project at the Houston Public Library, hosted by Houston's mayor, Sylvester Turner. Molly's stomach sank. "Oh, fuck," she thought. "This is going to be big."

A few hours later, she meets Michael Moritz and Ally Smither near the library entrance. The three are quiet as they filter into a meeting room full of elected and appointed officials. It is six days before Christmas. Molly has no idea what is about to be announced, but the sinking feeling in her stomach has lodged itself there.

Just after 11:00 A.M., Turner stands behind a wood lectern decorated with the city's seal, a navy circle wrapped around a Lone Star, a plow, and a "noble locomotive," which represents Houston's "spirit of progress." To his right hangs a seven-foot-tall map of Houston, the span of the North Houston Highway Improvement Project highlighted in green, purple, and orange segments. "I've always said, in its current state, I-45 is unsafe, it floods, and it doesn't meet our mobility needs," Turner says. "I believe this project could be transformative for the region and city of Houston if it was done right. It can be transformative, but it must be done right. Today, we are here to announce that the city and county have reached agreements with TxDOT." Those agreements are documented in two separate memorandums of understanding, one signed by the city and the other by the county. The city's MOU focuses on six areas: housing, drainage, transit, connectivity, parks, and right-of-way. "TxDOT has agreed to take only what is needed as this project moves forward," Turner says. There is no such thing as a perfect design, Turner says. But "on balance," he says, this was an "excellent project."

The mayor turns the lectern over to two county commissioners, who will vote to sign Harris County's agreement—which terminates the lawsuit against TxDOT—at their meeting on Thursday, three days from now. Commissioner Rodney Ellis has spent most of the past two years railing against TxDOT. The previous August, he biked to a press conference organized by Stop TxDOT I-45 outside Gallery Furniture, the showroom owned by Mattress Mack. "The idea that a freeway is going to mitigate flooding. Do you think we're drunk?" he cried, clad head to toe in Lycra. "The idea that a freeway is going to address climate change. Do you think we're stupid?" Since then, it has become clear: Rodney Ellis has no power. Despite Harris County's lawsuit—despite the federal government's intervention—TxDOT holds all the cards. TxDOT has the sovereign authority of eminent domain. It controls the state's transportation funding, and local leaders, it appears, are unwilling to let $9 billion walk away from Houston.

"I only want to single out in particular the activists," Ellis says. He looks at Molly and Ally and Michael, sitting with half a dozen other people who had been fighting the project for years. "The key for you activists to remember is in this business you ask for what you want and then you take as much as you can get. This lawsuit has gone to the end of its road and it was time to get that behind us," he says. "For decades, highways have torn apart Black and Hispanic communities, increasing wealth gaps and harming public health in communities all across America. We cannot afford to keep making the same mistakes. Harris County and Houston deserve twenty-first-century investments in infrastructure that take public safety and equity into account, as well as opportunity for everyone. The I-45 expansion as originally designed by TxDOT did not quite meet those needs." Ellis pointedly does not continue to say whether these agreements will create a highway that does, in fact, address those needs.

Commissioner Laura Ryan looks exhausted when she stands up to speak. Ryan, who lives in Cypress, an incorporated community in northwest Harris County, would be stepping down from the transportation commission the following month, after serving her six-year term. "Through patience, listening, compromise, and hard work, we have come to an agreement and stand ready to move forward together," she says. TxDOT would continue its "conversations" with the Federal Highway Administration, but she was confident a resolution was forthcoming. "This transportation infrastructure project can serve as a model for our country in a world where disagreements are often deal breakers."

Really? Molly thinks, sitting quietly in her chair. You think anyone is watching us to *learn*? She felt crazy, as if she were the only one who could see the emperor had no clothes. Whatever TxDOT had agreed to, it wasn't enforceable. The memorandums of understanding were nonbinding agreements, so anything TxDOT did was voluntary. And most of what Turner had outlined in his speech, TxDOT had already agreed to back in 2020, when it released its final environmental impact statement.

Finally, Turner pulls out a chair and sits at the table in the front of the room. With Laura Ryan on one side and Christian Menefee, the Harris County attorney, on the other, Turner signs and dates the city's memorandum of understanding.

After the press conference, Molly approached Menefee. "What happened?" she asked. "He was like, 'Look, the feds have no appetite for this.'" Molly was devastated. "It's so disappointing that they came down here and said all that stuff, acted like they care about Black and brown communities, acted like they care about carbon emissions," she says. "Come to find out that they were never going to do much." Molly, Ally, and Michael walked out of the library and into the cold. They got coffee, sitting in stunned silence.

"How are you feeling?" Michael asked Molly.

"Dead inside," she said. She went home and finished a few tasks, met friends for a beer, and went to bed. The next morning, she woke up vomiting, covered in a rash. "I had a complete nervous breakdown," she says. Michael, too, woke up sick; his throat was swollen and he was barely able to talk. "I think we were both processing those feelings physically," he says.

"This is how Texas is," Molly says. "You can't fight TxDOT."

It will be weeks before Molly can bring herself to read the memorandums of understanding. A few days after the press conference, she and Chloe drive up to their parents' house on Lake Conroe. In 2019, Mark and Judy had sold their house in The Woodlands and moved even farther up I-45 to Conroe, forty miles north of Houston. The Woodlands had gotten crowded, Mark said—too busy for them. And they thought their kids might come visit them more often if they were on the water. On one side of the living room, floor-to-ceiling windows looked out over the sparkling lake. Together, the family cooked and ate and played games and spun around the lake on Mark's boat.

Back in her house in midtown Houston, Molly finally opens the documents. Mostly, they focused on segments one and two—a twelve-mile stretch of I-45—and ignored segment three, which circled downtown. Segment three included I-10, the highway that would displace Modesti, as well as I-69, where the very threat of expansion had already forced people out of the Lofts at the Ballpark and Clayton Homes. TxDOT had made major commitments to increase drainage and flood mitigation as part of the highway design; that was good, Molly thought. It had agreed to accommodate more street crossings over the highway and to build wider sidewalks and bike lanes. It would "coordinate" with Houston's transit agency and include bus rapid transit lanes. But TxDOT had not agreed to make the highway narrower, committing only to "evaluate the design . . . to reduce the footprint and rebuild within the current footprint where feasible." Almost every commitment included the phrase "where feasible." Mostly, for Molly, the agreements were underwhelming.

In January, in a meeting with advocates, Menefee would say, "We had no leverage." After Harris County filed its lawsuit, he and people in his office met with the Federal Highway Administration, he said. "We talked to some pretty senior people. I believe the judge even had a conversation with the secretary himself. And all the feedback that we got, at every level of the federal government, was there is no appetite to substantially change the way this project looks. It was scoped more than a decade ago. What we need is for the city and the county to get on the same page. Under no circumstances will there be a push at the federal level to meaningfully change the way that the project looks."

19

FORWARD

Dallas
I-345

Growing up in South Dallas, Lakeem Wilson often wondered about the Forest Theater, the vast white building on Martin Luther King Jr. Boulevard. By the time he was in elementary school, the building had been abandoned for decades. "What was it like inside?" Lakeem wondered whenever they drove past, his face glued to the car window, peering up at its green tower. Lakeem listened, rapt, as his dad, Lockridge, talked about the theater—the movies he used to see when he was young, the music that wafted over the neighborhood. Lakeem loved his corner of South Dallas—there were always artists hanging around the house, musicians working on new tracks. "It had certain bad parts, but it was always out-shadowed by the richness of the culture," he says. He graduated from Lincoln High School in 2010 and left Dallas to study art at the University of Texas at Austin. In the fall of 2022, shortly after he moved back to Dallas, he heard about a nonprofit called Forest Forward that was trying to restore the Forest Theater.

Lakeem got in touch with LaSheryl Walker, a South Dallas resident and the director of community engagement at Forest Forward. He was a local artist, he said, a painter and graphic designer. Could he help? LaSheryl invited him to come see the theater. As they walked around inside, Lakeem marveled at the old murals propped against walls, massive plywood and plaster paintings by a local painter named Ellwood Blackman. Some were taller than Lakeem, depicting in chaotic gray scale the musicians who had once played at the Forest Theater—Stevie Ray Vaughan, Jimi Hendrix, Miles Davis. Lakeem felt a surge of hope as he walked through the theater and learned about the nonprofit's plans to bring it back to life. He came back to Dallas for his family, but also because he wanted to make art in the place he'd grown up. "I love the idea of things being restored, but I just want it to be community led, and the people that live here—I want them to be the ones that are able to restore it and bring it to life," he says. He started doing occasional projects for Forest Forward, helping design flyers and graphics for the nonprofit's website and social media.

For many years, on Martin Luther King Jr. Day, people from across southern Dallas gathered in front of the Forest Theater to watch as marching bands and cheer squads and local dignitaries perched on floats paraded down the boulevard named in honor of the civil rights leader. The parade had been canceled for two years during the pandemic. In 2023, LaSheryl and Elizabeth Wattley, Forest Forward's CEO, decided to host their biggest watch party yet, bringing South Dallas back to the Forest Theater. LaSheryl asked Lakeem to design a mural that community members could paint together along the front of the building's long facade.

Lakeem spent the weekend before Martin Luther King Jr. Day preparing the mural. Lockridge helped him sketch geometric block letters onto the plywood boards that covered the Forest Theater's front entrance, spelling "F O R E S T F O R W A R D" on a black background. On Sunday night, Lakeem's mom and girlfriend poured red, blue, and forest-green paint into plastic containers labeled with numbers. By Monday morning, two dozen kids and teenagers stand in front of the mural, filling in the enormous paint-by-numbers canvas Lakeem has conjured. He stands back, surveying their work, wearing paint-spattered jeans and a yellow bucket hat. "It's great," he says, a wide smile lodged on his face. "It's so great."

Behind him, a spectacle unfolds. The parking lot fills with people; a few dozen quickly becomes several hundred. People arrive with collapsible lawn chairs strung over their shoulders. Kids run in circles. Middle-aged women wearing matching sorority jackets gather for group photos. A DJ plays the "Cha Cha Slide," and a dozen women line up and follow the song's instructions—right foot, let's stomp, left foot, let's stomp. Elizabeth and her three-year-old daughter dance, wearing matching outfits: black leggings, a black Forest Forward T-shirt, and snug denim jackets. Smoke wafts from a grill where two men flip hamburgers and rotate hot dogs. An artist sits behind an easel and draws caricatures. Beside him, kids line up to get their faces painted.

Inside the Forest Theater, on a fold-out table covered in informational postcards, there are three plastic viewfinders fitted with circular film reels. It's quiet in here, cool and dim. I hold one of the viewfinders up and peer inside. The first image shows a rendering of the restored exterior of the Forest Theater, small figures sitting at patio tables under umbrellas, a tree-lined boulevard visible in the distance. *Click.* Inside the theater, the lobby is gold toned, a concession stand and lush movie theater carpet sweeping up toward the balcony. *Click.* A rooftop restaurant, backlit by a purple sunset, the downtown skyline in the distance. *Click.* A recording studio. *Click.* A classroom. *Click.* The signature green tower, illuminated and silhouetted against a navy sky. I lower the viewfinder and my eyes adjust back to the dim interior.

Just after 10:00 A.M., the parade begins with six dancers in gold-and-white tasseled leotards strutting down Martin Luther King Jr. Boulevard. The city had said it expected more than 150,000 people would congregate along the mile-long route. As the grand marshal drifts past on a purple-and-white float, waving at the crowd, two boys return the gesture with a perfect royal wave, palms cupped around an imaginary lightbulb. Eric Johnson, the mayor of Dallas, and his wife drift past. "He looked at me!" one of the boys cries. "He made eye contact. He knows me now!"

A woman approaches Lakeem and asks him if there's anywhere her son can paint. "Try the blues on the left side over here, the twos," he says. She comes back a minute later, reporting that her son is too short to reach the empty numbers. "Let me find a chair," Lakeem says. When he returns, I ask him if South Dallas has changed since he was a kid. He smiles his big, gap-toothed smile. A lot of it is the same, he says. But yeah, it's changed. "The major changes are the highways. They built the highway and now fifty years later, they're tearing the highway back down. I think it could be good or bad. I know in terms of real estate and things, it may increase the property values, which pushes some people out of the neighborhood. But at the same time, it can be an opportunity for people to take advantage of."

Over the past year, I watched as the S. M. Wright Freeway was dismantled. It was a shock, every time I visited South Dallas, to see the barrier removed, sight lines suddenly exposed across heaps of cinnamon-colored dirt. I watched as the concrete was hammered away, the steel munched and hauled. "This highway change is catalytic," Elizabeth tells me. "I totally think it's being underestimated. That is going to be storefront activity up and down the boulevard. It's slowing down to thirty-five miles per hour. We're going to have traffic lights. That is going to change the dynamics of retail and business and entrepreneurship in this community. We need to leverage the transformation rather than being consumed by it."

Lakeem had been living with his parents since he moved back to South Dallas but had just started to think about looking for his own place. He wanted to stay in the neighborhood; he was hoping to find work at the Forest Theater. "I'm definitely excited to be involved in that, knowing that it's my career path, knowing the impact that art had on me, on how you see your community, and how you express yourself through art," he says. "If I move too far, I'll just be trying to come back, driving all the time. I want to stay here and find something nearby and work from the inside out."

By 11:00 A.M., the parade is in full swing. It feels more like a rambunctious block party than a parade. Everybody waves to everybody. "I see you," the performers call to spectators. "How you doing?" they call back. Children yell and laugh and dance. As the Southern Methodist University marching band advances along the boulevard, clanging chimes and tooting horns, I walk east on Martin Luther King Jr. Boulevard, across the service road to the highway overpass. People sit in folding chairs on the sidewalk, in front of a metal guardrail. Before them, the South Oak Cliff High School cheerleaders smile and wave white pom-poms. Behind the spectators, below the elevated overpass, a wide dirt trench extends in either direction, a river of mottled amber streaked with blond and coffee, textured by the treads of construction tires. To the south, the rolling undulations of what was once a six-lane highway flatten. It's a national holiday and still the construction work continues. An empty dump truck trundles south, briefly disappearing under the overpass before it emerges on the other side and accelerates up a gentle slope. A few minutes later, when the truck returns north, the bed is full of mottled dirt.

I thread through the crowd and walk south along the service road on the east side of the former freeway. Where there was once a six-foot-tall concrete wall containing a hill of grass sloping up toward a guardrail—and beyond that, a highway—there is now only vacancy, a gentle depression in the earth across a channel of dirt. I continue walking south and stare across the expanse, looking at the houses that have been revealed, worse for the half a century that they have looked out over a highway. Cars are parked in driveways in the shade of sprawling oak trees. I pause. I am reminded of 1955, when a photographer leaned out of a plane and captured the South Central Expressway after hundreds of homes had been cleared to make way for a highway. *Click*. Homes on either side suddenly facing vacant space.

Click. The Forest Theater's bright green tower rising above it all. *Click.* And a wide dirt trench, scraped clean.

A few weeks later, on a Sunday evening in mid-February, alerts pop up on Patrick Kennedy's cellphone. A frantic city council member was trying to get ahold of him. "She called me in a panic, basically, and said, Hey, do you see what's on the consent agenda?" Patrick says. The city council was set to meet the following week, and the agenda had just been posted online. Tucked six pages into the consent agenda—which typically contained noncontroversial actions the council approved in bulk, without discussion—there was a resolution expressing the council's support for the "hybrid" option recommended by TxDOT for I-345, which included tearing down the elevated structure, lowering the highway into a trench, and connecting the existing street grid over it. The city council hadn't discussed the hybrid option since Ceason Clemens's presentation in the fall, and members remained split on TxDOT's proposed plan for the highway. To Patrick, it seemed as though someone had put the item on the consent agenda hoping to get it approved without attracting much attention.

Patrick started calling and texting other city council members. "Hey, have you noticed this?" he asked. Two days later, the resolution had been pulled from the agenda. Five of the council's fourteen members—including Paul Ridley, whose district spanned the highway, and Chad West, who represented the neighborhood of Oak Cliff in southern Dallas—wrote a memo asking the mayor to direct the city manager to commission an independent study looking at the removal of I-345 and to postpone any resolution supporting TxDOT's hybrid option until the study had been completed. There was no particular urgency to make a decision. TxDOT had recently spent $30 million retrofitting the existing highway, which extended the life of the structure by another twenty-five years.

And yet, by late May, the resolution supporting TxDOT's "hybrid" option was back on the city council's agenda. On a warm Wednesday morning, the council convenes in the angled concrete prow leaning over downtown Dallas. Thirteen people have signed up to testify on the resolution, every one in opposition to the council's support of TxDOT's plan. If I-345 was a connector to jobs, it was one "that chains the people of Dallas to serve in lands far away," says Hexel Colorado, the software developer turned freeway fighter who testified at the Texas Transportation Commission almost a year earlier. Removing I-345 would "align the incentives so that the people of Dallas support the people of Dallas, not people farther and farther away, out of city limits," he says.

DeDe Alexander, a Black woman in her thirties, asks the council members to consider "the past, present, and future of Dallas." TxDOT had failed "to take responsibility for past harms inflicted to Black neighborhoods," she says. Today, TxDOT was asking the city to "rubberstamp" the project without full consideration of its effects. And the future? "If we are going to build a city that's built on green principles, it cannot be a car dependent future." In early 2022, DeDe and her husband, Adam Lamont, had cofounded a group called Dallas Neighbors for Housing—a chapter of the national pro-housing group YIMBY Action—to advocate for more dense development across the city. They took up highway removal as their first campaign. The night before the council meeting, DeDe and Adam organized a community event at a restaurant on Ross Avenue, steps from the overpass where thousands of people had gathered to celebrate the opening of Central Expressway seventy years earlier. "Dallas is a very young city. Which means we have opportunity," said Jerry Hawkins, the executive director of Dallas Truth, Racial Healing, and Transformation, at the event. "The opportunity is to determine the identity we want as a city, and I believe . . . that is one of justice." As the summer sky faded to a brilliant blue, the group walked under I-345 to hang a large canvas painting of a bustling city street, reading COMMUNITY NOT COMMUTE.

Back at city hall, council member Omar Narvaez introduces the resolution with an amendment: The city's approval would be contingent upon TxDOT investigating alternate sources of funding, including the Reconnecting Communities Pilot Program, "for studies regarding alternate design options, including . . . new options for the future of IH 345," he says. The city reserved the right to "withdraw its support of the refined hybrid option." But such a reversal would be costly. "Should the council choose to change their mind, we would seek to be reimbursed the engineering dollars that we're about to expend," Ceason says. "We're anticipating this next step to cost close to $20 million."

"I think that would be a stretch to say I could find $20 million . . . in our general fund budget," responds assistant city manager Robert Perez. "I think that would be a pretty high bar for us."

"I've made it very clear that I'm not in favor of a trench," says council member Chad West, the first council member to speak. "I'm not in favor of an elevated freeway. I'm not in favor of a highway of any type continuing to divide the neighborhoods of Deep Ellum and downtown. A boulevard system . . . would provide for a lot more economic development, housing, restitching our neighborhoods, things that we all say are a priority to us, not just moving cars." But he has exhausted all efforts to delay the inevitable. "I'm going to grudgingly support this amendment today," he says.

His grudging support will prove prophetic. After an hour of discussion, the council unanimously votes to support the resolution. "After a dozen years the push to remove I-345 appears dead," writes Matt Goodman, a local journalist who had covered the campaign for removal for the better part of a decade. With the council's support, TxDOT could request funding from the state and proceed with engineering plans.

When Patrick started his campaign to tear down I-345, he wanted to have a citywide conversation about the future of Dallas. "What does Dallas want to be when it grows up?" he asks. Dallas was still reckoning with all that had ailed it during the twentieth century—

the embedded racism, the segregation and sprawl, the vast inequality. He wanted to have that debate. "People apparently love their trucks and love their cars. Well, let's test that, because I don't necessarily believe that," Patrick says. "I think it's more Stockholm syndrome, that people love it because they have to and are held captive by it. We're trying to change the bones of the city and rearrange how the land uses organize themselves around those bones," he says. Changing the bones of a city like Dallas—a place that had calcified around the automobile—was a long fight. Patrick knew that when he got into it. He wasn't going anywhere.

20

FIGHT

It's dark by the time Jaime Cano arrives at the Whip In with his wife and two daughters on a Saturday evening in January. An immigrant from India named Amrit Topiwala had opened the Whip In as a convenience store decades earlier—the name referencing the maneuver cars made as they entered the parking lot from the fast-flowing frontage road—but the space had gradually morphed into a full-service restaurant and beer garden. Tonight, a hundred-odd people have gathered under a carport strung with globe lights. As Jaime winds through this crowd of strangers, his daughters—nine-year-old twins—trail behind him, their eyes wide. He spots Luke Metzger, the executive director of Environment Texas, an advocacy group and co-plaintiff on Rethink35's lawsuit against TxDOT. Both of Luke's older kids graduated from Escuelita; now his youngest daughter spends her days in the copper room with Ms. Delvis. As they chat, Jaime's daughters wriggle around him, climbing on his back and dancing around his legs. Behind them—past a galvanized metal garden bed and a few locked-up bikes, beyond a narrow strip of parking and a row of Rethink35 yard signs—is the highway, cars streaming through the cement river.

A few weeks earlier, TxDOT had published its initial environmen-

tal review for the I-35 Capital Express Central project, as required by
the National Environmental Policy Act. Before the draft environmental impact statement was released, all anyone had seen were
TxDOT's rough schematics, dotted lines traversing parcels on a map.
Nothing was final, the agency insisted; all alternatives were being
considered. In December 2022, TxDOT finally identified its preferred design for the highway and all the property that would be required to build it. Along eight miles, the highway would consume
almost forty-two acres of land and displace 107 homes and businesses, including Escuelita del Alma. (The Whip In had narrowly
been spared.)

"Seeing our name in there really made it real," Jaime tells me now.
It had all gotten much more real recently—the parents asking what
would happen to the school, the teachers worrying about losing
their jobs. Jaime learned about other businesses that would be displaced, including a barber who didn't know his shop was in the path
of the expansion until a local TV reporter stopped by to talk to him.
"I wouldn't know what to do or where we would restart at," the barber said. Jaime read the story and felt outraged on the man's behalf.
"The impact statement doesn't include that part of it, that sense of
instability that people feel, especially when they had no idea," Jaime
says. "Part of me wants more details, and at the same time I don't
want to get too bogged down. It's really daunting. It just sometimes
makes it feel a little more insurmountable."

TxDOT hadn't yet made Richard Linklater an offer for the building; all he knew was that it was coming. "It will not be the same if La
Escuelita has to go way out of town to find something similar they
can afford," he wrote to me in an email. "It only seems fair that
TxDOT should accommodate those they are displacing by making
sure they can find the same size space that works for the particulars
of what they need for their business."

The environmental impact statement, which spanned more than
seven thousand pages, included a greenhouse gas analysis specific to
the I-35 expansion. In 2017, when TxDOT looked at the greenhouse

gas emissions resulting from the North Houston Highway Improvement Project, it concluded that all alternatives "would reduce GHG emissions by improving traffic mobility, which results in less idling, increased travel speeds, and a reduction in vehicle fuel usage." (Increased speeds reduce emissions because cars use fuel more efficiently the faster they go—up until thirty miles an hour, when that benefit flattens.) No mention was made of induced travel. In the I-35 study, TxDOT acknowledged that "transportation related GHG emissions may contribute to climate change" and that expanding I-35 might increase those emissions. According to TxDOT's analysis, the induced travel generated by expanding I-35 through central Austin would add nearly 500,000 metric tons of carbon dioxide equivalent emissions by 2050. But an independent analysis by a sustainability nonprofit called Rocky Mountain Institute found that expanding I-35 through central Austin would generate almost eight times that, as much as 3.9 million metric tons of carbon dioxide equivalent by 2050.

TxDOT was sure of one thing: Congestion would get worse. According to TxDOT, average daily traffic on I-35 would climb from 207,000 cars in 2019 to 304,000 in 2045, an increase of 47 percent. Driving eight miles south to north across Austin "should take approximately 8 minutes," TxDOT reported. In 2019, that same trip at evening rush hour consumed 32 minutes. By 2045, according to TxDOT's travel demand models, traveling eight miles up I-35 would take 223 minutes, or 3.7 hours.

It was ludicrous—the idea that you "should" be able to get from one side of the country's fastest-growing city to the other in only eight minutes. But more than that, the models were simply wrong. No rational person would spend 223 minutes driving on a highway to travel eight miles. Imagine getting in your car, pulling up your Google map, and being told your eight-mile trip would take 3.7 hours. You simply wouldn't go. You would take a different route or travel by other modes. Most people could *walk* eight miles in less than 3.7 hours. What's more, the projected travel demand assumed a future in which a bigger highway had already been built. Since 2000, daily traffic counts on I-35 through central Austin had hovered around 200,000 cars; that's all that could fit on the highway. The only way to accommodate 304,000 cars was to build a bigger highway.

The models did not predict the future. They simply justified the expansion. And yet TxDOT's authority was absolute. What layperson could challenge the premise of a bad traffic model, especially when it was so meticulously documented and professionally presented?

Stopping the I-35 expansion would be a tough fight, says Metzger, when he takes the stage. "But I think we have a lot of reasons to have hope. If you just look all around the country, we're seeing victories against freeway expansions just like this." In May 2022, the Los Angeles County Metropolitan Transportation Authority unanimously voted to cancel a $6 billion expansion planned for the 710 Freeway after nearby communities—which were predominantly Black and Hispanic—organized and fought back. That same month, facing pressure from environmental groups, the Colorado Department of Transportation canceled a plan to spend $1.5 billion to expand I-25 through the heart of Denver. In Portland, after the Oregon Department of Transportation, or ODOT, allocated $1.2 billion to expand a 1.8-mile stretch of Interstate 5 through a historically Black neighborhood, a group of teenagers started protesting outside the state agency's downtown headquarters. "People don't think of ODOT as a villain in the climate crisis, but they don't realize that 40 percent of our state's carbon emissions come from transportation, come from the freeways that ODOT is trying to expand," said Adah Crandall, a fifteen-year-old organizer. Starting in April 2021, every other Wednesday evening, Adah and a group of friends gathered outside ODOT's headquarters. They scrawled chalk messages on sidewalks, pinned banners to the side of the building, and held signs: CLIMATE LEADERS DON'T WIDEN FREEWAYS. Ten months after Youth vs. ODOT began its campaign, the Federal Highway Administration rescinded its approval of the project and requested that the state redo its environmental study of the project.

"These things didn't just happen," Metzger says. "They happened because of fierce opposition from freeway fighters like you, people all around the country that are saying 'hell no.'"

Adam Greenfield stands midway through the crowd, arms crossed over his white Rethink35 T-shirt and corduroy blazer, a grin spread across his face. Two years ago, Adam was an activist sitting on a hill, looking at a highway. Since then, hundreds of people have joined Rethink35's fight. More than five thousand people have subscribed to the organization's mailing list. A group of students and faculty at the University of Texas at Austin had started their own Rethink35 chapter. "On a climate change front, we want less cars," a junior at UT said, succinctly, when the student group launched. People in Round Rock and San Marcos, suburbs north and south of Austin, had started their own Rethink35 chapters, disproving the assumption that all suburbanites wanted was bigger highways.

Leading up to Austin's mayoral and council elections in November, Rethink35 held candidate forums, pressing those running for office to answer whether they supported the expansion. The mayoral race had been incredibly close, with the former state representative Celia Israel in a runoff against the former state senator—and former Austin mayor—Kirk Watson. "The current plan to expand I-35 is unacceptable," Israel told Rethink35. "Adding more lanes won't ease congestion, it will only increase reliance on cars." Watson, on the other hand, supported TxDOT's vision; he was, in many ways, the author of it. When asked about the highway expansion before the election, Watson said, "Whether you like cars or not, people are driving cars and this would make that road operate better." In a low-turnout December runoff, Watson beat Israel by just over nine hundred votes.

"Hola, hola, buenas noches, como están?" Jaime says, when he is invited to the stage. "Thank you, all of you, for the work that you're doing. I'm going to try really hard not to cry. I'm very passionate about the work that I do," he says. Running a preschool was demanding work, all consuming, and it was easy to ignore the existential challenge that Escuelita faced. "But I sleep better at night knowing that at the very least there are people like you who are fighting for us." He pauses, his voice catching, tears welling up. "And fighting for all the local businesses and residents that are going to be displaced by this expansion. Literally, you are fighting for the littles." A ripple of laughter passes through the crowd, relieving some pent-up pressure. Several people watching cry along with Jaime. "You're fighting for the littles at the most rapid developmental stage in a human's life, zero to five years old. They need a place to go and to learn. Families need a place that they can trust and that they can count on to send their most precious family members."

Jaime, too, is a different person than he was a year and a half ago, when he spoke in the parking lot of Stars Cafe, just south of Escuel-ita, below the thundering upper decks of I-35. There, he read from a script. He was professional. He was brief. Tonight, he is emotional; he is raw. His voice wavers and stumbles as he fights back tears. "Seeing all of you here today inspires me in a way that I haven't felt in a real long time," he says. "I want to thank you all from the bottom of my heart. The current proposal is going to be devastating to not just properties and people and the environment in Austin, but to the spirit of what Austin is supposed to be. I'm from here. And I've seen it change in very positive and very negative ways. But Austin and the spirit of Austin, the soul of Austin, has really held strong, but these are the kind of things that threaten that very soul and that very spirit."

Throughout the fall, Jaime and Dina kept circling back to the same question: Where would Escuelita go? They didn't have an an-swer, but they knew that this was the last move the school could survive. As 2023 began, they started looking for a building to buy, searching for a space that could safely contain all the little people and little things they were responsible for. Escuelita was Dina's life's work, and Jaime couldn't let it go without a fight. "I just want you to keep in mind that when I go back to work, when I answer questions from parents about what's going to happen to our school, when I talk to the teachers about what's going to happen to our jobs, what's going to happen to these kids, where are they going to go?" Jaime says. "I can tell them that the future is uncertain but there are people out there fighting for you and fighting to make sure that at the very least there is a chance that we get to stay open and that we get to continue serving these families. I get to do that because of you guys. And that's so important, because where there is organizing, where there are people that care, there is hope."

Throughout the winter and spring, TxDOT barrels forward with its expansion plan. In February, the agency holds a public hearing on the draft environmental impact statement at a community center in East Austin. Yet again, schematics sprawl across tables and poster boards lean on easels. Outside, nearly thirty Rethink35 protesters gather, chanting and holding signs.

At the public hearing, TxDOT announces that it will release a combined final environmental impact statement and record of decision by late summer 2023. A record of decision would allow the project to begin construction. The sooner it broke ground, the harder it would be to stop.

Later that month, Austin's city council spends an afternoon considering a resolution containing a list of demands to send to TxDOT, including more east-west crossings over the highway and a smaller footprint to "minimize the number of properties impacted by eminent domain." Although many council members opposed the expansion, the resolution does not explicitly condemn it. Elected officials knew they had little leverage. The project was already funded, and so, unlike in Dallas, TxDOT did not need Austin's approval to move forward. "I'm conscious of the power dynamics between the state and the city," says the council member Chito Vela, whose district hugged the highway. When the city council opens the meeting to public comment, Heyden Black Walker reminds the council that Reconnect Austin's vision—which would free up dozens of acres of land by depressing and covering the highway—represented more than $3 billion in development opportunity, which would generate hundreds of millions of dollars in tax revenue for the city. Sinclair Black follows his daughter, trenchant as always. Then, finally, Addie Walker—Heyden's daughter—speaks. Two years earlier, Addie had graduated from Washington University in St. Louis with a degree in public health and moved back to Austin to join the Reconnect Austin team. Now the fight against I-35 spans three generations. She is soft-spoken like her mother, but clear and direct. "I am twenty-four years old, and once constructed, this project will stand for my entire life," she says. "We cannot allow TxDOT to bulldoze more of our city and further separate downtown from East Austin. I live . . . east of I-35, and I don't want to be separated from my family for the rest of my life."

More than three dozen people testify. Not a single person speaks in favor of expanding I-35.

Kirk Watson, Austin's new mayor, is the only representative to vote against the resolution. He had joined the Climate Mayors coalition when he took office, committing to climate leadership through "meaningful action." That commitment did not, apparently, extend to the highway. I-35 was his legacy; he'd fought for years to get the expansion funded. Earlier that week, he posted a note to his colleagues on the council's message board in response to the proposed resolution. TxDOT's design "allows for the opportunity" to build caps, or deck plazas, over the highway, but the city had to pay for the structural supports as well as the caps at a total estimated cost of $800 million. For TxDOT to include the caps in its construction documents, Austin would have to determine the source of the structural support funding—an estimated $325 million—by the fall of 2024. "I question the feasibility of demanding TxDOT design the project as dictated by the resolution," Watson wrote, "given that we have yet to identify a funding source for the caps alone and likely won't be able to meet the fall 2024 deadline."

It felt like a bait and switch. TxDOT had sold its highway expansion with the promise of a reconnected Austin. But unless the city council could conjure $325 million in less than two years, the city would end up with nothing but a wider gash cutting across it.

At the end of February, the Department of Transportation announces the forty-five projects awarded a total of $185 million through the inaugural Reconnecting Communities program. "Transportation should connect, not divide, people and communities," Secretary Pete Buttigieg says when he announces the grant recipients. The City of Austin won a $1.1 million planning grant for its Our Future 35 program. Funds would be used to "identify affordable housing, anti-displacement and business support strategies for neighborhoods surrounding new freeway caps," the city wrote in its application. It did not mention the affordable homes and local businesses that would be displaced to build the freeway that would hold those caps.

In March, TxDOT holds a groundbreaking ceremony to mark the start of construction on the I-35 Capital Express North project. Outside TxDOT's Austin district headquarters on the I-35 frontage road—in the shadow of a soaring on-ramp shuttling cars from U.S. 183 onto the northbound interstate—dignitaries in suits gather yet again to toss ceremonial dirt. Two dozen Rethink35 protesters stand on the periphery of the event, chanting and waving signs. "Today, we're going to celebrate the second part of this major project," Bugg says. Adam and other protesters boo and yell, their voices audible under the white canopy tent. "We're going to have the groundbreaking in just a few moments on the I-35 Capital Express North section. Then the next groundbreaking is going to be, what, in a couple of weeks, Tucker?" he says, turning to Tucker Ferguson, TxDOT's Austin district engineer. Tucker shakes his head, smiling, and the crowd laughs. "Tucker's working hard," Bugg says. The next groundbreaking would be the central portion of the I-35 Capital Express project, which had yet to be formally approved. A confetti cannon showers Bugg and other officials with colorful paper as they toss shovelfuls of beige sand onto the grass below.

In early February, Rethink35 organized a Youth Day of Action, a coordinated protest at colleges and high schools across Austin. More than a hundred students and teachers congregated in campus quads. They held up banners and talked about climate change and air pollution, speaking with a moral authority distinct to young people living on a planet polluted by their elders. "It'll be our lives and the generations after us who will be most affected by the decisions made today," one student said. On the University of Texas at Austin campus, a student a cappella group called Beauties and the Beat sang "Highway to Hell."

Judah Rice grew up in Austin and spent most of his life—all nineteen years of it—living within a mile of I-35. "When I step out on my porch at night, I can hear the roar of the freeway," he said at the UT rally. TxDOT was counting on young people not to recognize that "the playbook that they are pulling on us right now . . . is the same playbook that highway expansion advocates have been using for decades upon decades." Austin was in many ways a success story, Judah said. In the 1960s, there were plans to thread highways across the city—along Town Lake, up Guadalupe, across Martin Luther King Jr. Boulevard. "Imagine what it would be like going to UT with MLK as a freeway," he said. "They told us that it had to be done. We had to fix traffic." But public outcry had killed those highways; they had been erased from the map. "These things are not inevitable. We can stop these things if we actually believe that we can."

People in the streets had stopped highways before. Sixty years earlier, on Long Island, Judah's grandmother had pointed a shotgun at Robert Moses when he showed up on her doorstep, trying to buy her land to build a freeway through her backyard. "You know what? She won." Despite what TxDOT wanted Austinites to believe, the expansion of I-35 was not inevitable. Even now, TxDOT could still be stopped.

21

RESOLUTION

Houston
I-45

In late February 2023, Molly Cook wakes up early and drives west. Over the past year, she's made a habit of driving the two and a half hours to Austin every month to give a three-minute testimony at the Texas Transportation Commission. Today, the commission will begin discussion on its 2024 Unified Transportation Program, which would allocate a proposed $100 billion for road construction and congestion relief over the coming decade. "This 10-year plan will further boost our economy and keep Texas the economic juggernaut of the nation," said Governor Greg Abbott. "Together we are working to ensure that Texas remains the premier destination for people and businesses." Texas was flush with cash. Over the coming five years, it would receive nearly $27 billion for highways from the Infrastructure Investment and Jobs Act.

After the commission reviews the UTP's various funding categories, J. Bruce Bugg, Jr., calls up Brandye Hendrickson, TxDOT's deputy executive director, for a status update on the North Houston Highway Improvement Project. It had been almost two years since TxDOT had received a letter from the Federal Highway Administration, pausing the project. "We have been at parade rest, if you will, ever since," Bugg says. "So how do we get out of parade rest and get back into building this very important project?"

TxDOT had some "really great movement" last year, with agreements signed by Harris County and the City of Houston, both of which demonstrated community support for the project, Hendrickson says. Meanwhile, TxDOT had worked "tirelessly" with FHWA to address Title VI concerns. "We believe that we have come to terms with a voluntary resolution agreement," she says. That agreement had been sent to Secretary Buttigieg's office and awaited his signature. "We are cautiously optimistic that the project, given its importance and the support of the local community, will be able to move forward," she says.

"I'm just trying to understand where things really stand on that, because we have a lot of money set aside," Bugg says. As the delay dragged on, the estimated cost had ballooned to $9.7 billion. He unfolds a printout of TxDOT's budget. "I show on our internal records that the earliest start date would be in 2027, which is of course four years from now. And the latest start date on certain segments goes as far out as 2036," he says.

A month earlier, the Houston-Galveston Area Council had voted to delay funding for a portion of segment three until 2027, citing "readiness issues." Because of the federal pause, the TxDOT district engineer Eliza Paul reported to the council, "there is no way we can get that project ready by 2024." Segment three, which circled downtown, was always intended to be the first stretch of the massive project to break ground. In 2018, when Houston joined a North American bid to host the 2026 World Cup, rebuilding the downtown loop was seen as a foregone conclusion. Officials planned a massive improvement campaign for Houston in advance of the games, "potentially kick-starting a project to develop a parklike space atop a buried freeway east of the central business district," reported the *Houston Chronicle* in 2018. But after FHWA paused the project, even the NHHIP's most vocal proponents were forced to admit they didn't want downtown torn up just as people from across the world descended upon Houston. In June 2022, the president and CEO of Central Houston Inc., a downtown business alliance, wrote to Paul, the district engineer. "After we collectively celebrated the winning of a World Cup bid, the reality of preparing our city to host the world started to sink in," he wrote. "In particular, NHHIP segments 3A & 3B, which could still be under construction in 2026 given the current, uncertain timeline. Houston's World Cup committee chair, Chris Canetti, is copied on this correspondence. In speaking this morning, he agreed that it is in our collective best interest to open a discussion about the potential timing of the project in coordination with the games." And so TxDOT had agreed to delay the project even further.

Now Bugg would like to suggest that TxDOT consider reallocating the nearly $10 billion dedicated to the NHHIP to other projects in the Houston region. There were plenty of road projects in the pipeline, projects that had been held up because of a lack of funding, he says. In the meantime, the money set aside for the NHHIP was "getting whittled away by inflation"—the cost of highway construction had increased by 50 percent since 2020. "We might as well move it over and get these other important projects built," he says. He asks Hendrickson to consider whether this idea has merit and to report back.

Commissioner Alvin New jumps in with a point of clarification. "I want to be sure that we're sending a very clear message," he says. The NHHIP was still going to happen. This was simply a financial maneuver. Yes, Bugg says, exactly. "This is not a choice of ice cream or cake. We want to give the Houston area ice cream and cake," he says. "But the timing is, the cake is not coming out of the oven for a long, long time. We might as well serve them ice cream in the meantime."

Sitting in the audience, Molly is quietly—cautiously—thrilled. She was "nervous as a cat" about the pending agreement coming from FHWA, which had the final say on what would happen to those living in the proposed right-of-way of the expansion. The threat of displacement still loomed. But for now, they were safe. No one could be compelled to move until at least 2027. The extra time gave people options, the ability to plan. Modesti could stay in her house for another four years if she wanted to. Tenants at Kelly Village, the public housing complex across the highway from Clayton Homes, could stay in their units. Because the downtown loop would be the first segment to break ground, the stretch of I-45 running alongside Elda and Jesus's home, north of downtown, had been delayed even longer.

A lot could happen in four years.

A coalition of nonprofits and community groups fighting the I-45 expansion had recently received a $180,000 grant from the Energy Foundation, a philanthropic group focused on tackling the climate crisis. It wasn't much, but it was a whole lot more than the nothing they had been working with. They started meeting regularly, calling themselves the I-45 Coalition. The fight felt more professional now. Ally Smither, the opera singer who took over Stop TxDOT I-45's communications, had been hired by Air Alliance Houston, a nonprofit that had opposed the highway expansion for years, and was now getting paid to help organize the resistance. Molly felt fired up, energized.

Bugg opens the agenda item to public comment. Two people have signed up to testify: Molly and Jay Blazek Crossley. When her name is called, Molly walks briskly to the lectern. She is polite and chipper as she offers comments on transportation financing, safety in urban areas, and outcome-driven metrics. "And then just a friendly reminder," she says, "that more lanes mean more traffic and they have for over ninety years. These projects don't work to relieve traffic congestion in any sort of meaningful long term. We've got to have buses. We've got to take people out of their cars."

Two weeks later, in mid-March, Shailen Bhatt arrives in Houston. In December, Congress confirmed Bhatt as the administrator of the

Federal Highway Administration, replacing Stephanie Pollack, who had been acting administrator since shortly after President Biden took office. Congresswoman Sheila Jackson Lee gathers a group of community leaders and elected officials on a METRO bus just as she had done when Pollack visited Houston. The tour of affected places follows the same itinerary as a year and a half before. But now many of these places are abandoned. At the Lofts at the Ballpark, the group pauses to survey the vacant block on St. Emanuel and peer into the uninhabited buildings behind it. At Clayton Homes, Bhatt walks over shards of glass scattered from broken windows. Doors are kicked in, blinds bent and broken, the public housing complex deserted. At Greater Mount Olive Baptist Church in Independence Heights, the group congregates in the cracked parking lot on the highway frontage road before walking inside the historic church. "Do you still come here to worship?" Jackson Lee asks the pastor, a Black man in his sixties, as the group filters into the nave, standing among wooden pews.

"Oh yes, ma'am. We come in every Sunday," he says.

"Good. Keep it up," she says. "How old is this church, Pastor?"

"A hundred and . . ." He pauses. He's not sure. After a quick investigation, he reports the church was founded in 1903. "A hundred and twenty years," he says. After the original structure was damaged by Hurricane Ike in 2008, the congregation had rebuilt the church in the same spot. "That's why we don't want the structure to go down," Jackson Lee says.

A few hours later, the group reconvenes in the Mickey Leland Federal Building, a towering glass structure adjacent to I-45, where a group of local reporters have set up cameras and microphones. Jackson Lee stands behind a lectern surrounded by more than a dozen people, including Bhatt; Eliza Paul, TxDOT's Houston district engineer; Sylvester Turner, Houston's mayor; and Susan Graham, who wears an oversized red Stop TxDOT I-45 T-shirt. They stand shoulder to shoulder, uncomfortably close, an awkward alliance of opposing parties. "Before I bring up Administrator Bhatt, I just want to show the legitimacy of the document I have here," Jackson Lee says, shuffling through a stapled leaflet of white paper. "Let's find the page. It's even blue ink, I've got the blue ink," she says, holding up a page with two signatures in looping cursive: Shailen Bhatt and Marc Williams, TxDOT's executive director. "Blue ink, real. And it covers all our concerns," she says.

On March 7, almost two years to the day after the Federal Highway Administration had paused the North Houston Highway Improvement Project, the federal agency had issued a voluntary resolution agreement, or VRA, which ended the civil rights investigation and allowed TxDOT to proceed with planning and construction. No finding was made as to whether TxDOT had violated the Civil Rights Act. Instead, the agreement committed the state agency to mitigation measures, focusing on concerns that had been identified by complainants, community members, and elected officials. "I think sometimes in transportation people who make decisions aren't the ones who are the most vulnerable," Bhatt says, after Jackson Lee invites him to the microphone. "I think what we want to see is that transportation is something we do *with* people, not *to* people." Bhatt was hopeful that the agreement signed by FHWA and TxDOT would produce a better outcome for Houston. "I probably won't be the federal highway administrator when this is done, but hopefully I get to come back and see a project that has been built with the community and that all boats have been lifted by this investment," he says.

Michael Moritz, who is tall and lanky, hovers in the back row, his face stoic. Along with other advocates and community members, he had been given a preview of the voluntary resolution agreement the day before it was released, in a last-minute Zoom meeting convened by FHWA. He was disappointed. So was Molly. Advocates across Houston insisted the agreement didn't address the underlying civil rights issues complainants had raised, because the project's negative consequences would still be predominantly borne by communities of color.

The federal government would be monitoring the project "very closely," Nichole McWhorter, the head of FHWA's Title VI team, said in the call with advocates. TxDOT had committed to convene biannual public meetings during planning and construction, submit regular progress reports to FHWA, and expand language access for people who didn't speak English. The state would spend an additional $3 million for affordable housing, an 11 percent increase from the $27 million it had already budgeted. It would allocate another $1.5 million to build new parks and trails. TxDOT would design and build four structural caps stretching over the expanded highways. It would install three air monitors along the twenty-eight-mile project. It would complete detailed drainage studies, using models developed by the National Oceanic and Atmospheric Administration. And, finally, TxDOT would evaluate "reasonable opportunities to reduce the project footprint in ways that would not compromise the integrity and functionality of the purpose and need of the Project."

Should TxDOT fail to comply with the agreement, McWhorter said, FHWA could issue a letter of finding, affirming that the state agency had violated the Civil Rights Act. It could withhold federal highway funds. Or it could refer the project to the Department of Justice to litigate enforcement.

Modesti Cooper listened as McWhorter and her colleagues outlined the mitigation measures TxDOT had agreed to, including increased funding for affordable housing. "You mentioned the affordable housing, but what about the displaced residents and business owners?" she asked, according to notes of the meeting. "No one has given us the opportunity to give input for a redesign. How does this cover the people who actually filed the Title VI complaints?" It felt as if TxDOT had gotten its way and FHWA would just be monitoring that way, she said. For more than a decade, TxDOT had insisted that fixing the highway required adding capacity: "Traffic congestion, which is measured by traffic volume and roadway capacity, will increase if no improvements are made." An expanded footprint naturally followed. Without challenging the premise of the project—without addressing the basic phenomenon of induced demand—the footprint was unlikely to change.

The Federal Highway Administration maintained it had limited authority to make changes to the project. "This is a formula project for a NEPA Assignment state, meaning we, the federal government— Secretary Buttigieg, the president—we don't pick the project, we don't design the project," Andrew Rogers, the deputy administrator of the Federal Highway Administration, told me. Rogers had been the agency's chief counsel during the investigation, leading the team that negotiated with TxDOT. The footprint of the highway had been determined during the NEPA process, he said. To significantly reduce that footprint, TxDOT would have had to produce a new or supplemental environmental impact statement and put it out for public comment. TxDOT was unwilling to do that, he said. "Now, it doesn't mean that there wasn't a strong appetite on the part of federal highways, city, and county to press them," he says. "And we did, but they were unwilling to, as a matter of agreement in a VRA, to effectively say our NEPA process didn't work, we're going to go back to the start. . . . So instead, what we asked them to do is to evaluate every reasonable opportunity to reduce the project footprint."

The same day that the voluntary resolution agreement was announced, FHWA sent a letter to Marc Williams, TxDOT's executive director. FHWA had concluded its audit of the state's compliance with the NEPA Assignment program. The FHWA team had not identified any issues with the North Houston Highway Improvement Project's environmental review process. Instead, the federal agency commended TxDOT "for working with Harris County and the City of Houston to resolve their concerns about many elements of this project as it proceeds to construction."

22

PROXIMITY

Beth Osborne stands on a stage in a seminar room in George Mason University's Van Metre Hall. Several hundred planners and mobility advocates have gathered in Washington, D.C., for Transportation for America's annual TransportationCamp, an informal "unconference" in which participants create the program. Over the previous hour, Beth watched as people pasted giant Post-it notes to an empty wall, scrawling panel titles in Sharpie marker: "Making Shared Mobility Bipartisan" and "Beyond Electric Vehicles."

"These are really, I'll say, exciting times," Beth says now. "They're a little nerve-racking times, too. We have a lot of funding from the federal government rolling down into states and into localities to build a lot of new transportation. But there aren't a lot of parameters on it. So it's going to be up to us, transportation professionals, advocates, and leaders, to make sure that funding is spent in a different way than it's been spent before."

Despite the Biden administration's commitment to sustainability and equity, most of the transportation money in the Infrastructure Investment and Jobs Act would be spent as it always had been: enabling seamless car travel. Only now, those cars would be electric cars; Biden wanted two-thirds of the cars sold in the United States to be electric by 2032 and had committed $7.5 billion to build out a national network of EV chargers. In the spring of 2021, Biden was delighted to test-drive a Ford F-150 Lightning, an electric pickup, in

Dearborn, Michigan, exclaiming as he accelerated away from a gaggle of reporters, "This sucker is quick!" American automobile companies were rolling out massive electric vehicles to attract American consumers, erasing some of the environmental advantages gained by electrification. The electric GMC Hummer, for example, weighed 9,063 pounds and generated more emissions per mile than a gas-powered Chevrolet Malibu.

Electric vehicles offer a false promise—that we can all keep driving as much as we want and still reduce carbon emissions. Electric vehicles fit into the built environment we already have, and so they seem like an easy fix—yet another consumer product that will save the planet. But unlike lightbulbs or toilet paper, people don't buy new cars very often: The average car on the road is twelve years old. In scientific terms, "the fleet turnover time is a bottleneck to transport decarbonization," writes Costa Samaras, a professor of civil and environmental engineering at Carnegie Mellon University.

To be sure, electrification is important. Reducing emissions from passenger cars and trucks requires both getting more electric vehicles on the road and decarbonizing the electric grid that powers those vehicles. "But there's a third leg in the carbon stool: how much we drive," Samaras writes. Twenty years ago, drivers in the United States covered roughly 2.75 trillion miles a year. Today, we drive 3.25 trillion miles annually. It doesn't do a whole lot to reduce the emissions of each mile traveled in a car if people have to drive more miles to get where they are going. This is precisely what happened in the past: Since 1990, even as fuel efficiency increased by 18 percent, total vehicle emissions jumped by 22 percent.

To meet the carbon reduction targets set by the Intergovernmental Panel on Climate Change, a report by the Rocky Mountain Institute found, the transportation sector needs to reduce carbon emissions by 45 percent by 2030. To do that, we need to put seventy million electric vehicles on the road in less than a decade *and* reduce driving by 20 percent. To reduce driving, we need to reduce distances—between homes and schools, apartments and jobs, bus stops and grocery stores. Mostly, that requires longer-term changes in land use and planning. Before Austin's land development code rewrite was halted by a dozen homeowners, city staff had estimated that allowing for more dense residential development throughout the city would reduce vehicle miles traveled by 23 percent.

The easiest—and quickest—way to reduce driving and lower emissions is to stop widening urban highways. A 2021 study by the Georgetown Climate Center and Rocky Mountain Institute found that the funding enabled by the Infrastructure Investment and Jobs Act could increase greenhouse gas emissions by nearly 2 percent over the coming decade if states continued to prioritize highway expansion over maintenance. Another study looked at the ten states with the highest gasoline consumption and found that "minimizing further highway expansion was the most important lever to avoid putting upward pressure on transportation emissions." Texas had the potential to add more greenhouse gas emissions than any other state through highway expansion.

Back at TransportationCamp, in a classroom on the second floor of Van Metre Hall, Martha Roskowski, a mobility consultant from Colorado, stands next to Beth in front of a white projector screen. Martha met Beth two decades earlier, when Beth was a legislative staffer and Martha was an advocate campaigning for more federal funding for bike infrastructure. Over the years, their paths kept crossing. A few months earlier, they had sent out a survey to freeway fighters across the country. They wanted to know: What does the freeway fighting movement look like today? And what did people need?

Fifty-eight freeway fighters from twenty-seven states responded to their survey, Martha reports. Some were part of nonprofit organizations; some were individuals motivated to spend their free time fighting highways. Nearly 60 percent of campaigns were unfunded. "This is a hugely volunteer-driven effort," Martha says. The screen fills with a painting of a massive, armored Goliath towering over a thin-skinned David. "I went searching for the best David and Goliath picture I could find," she says. But now, looking up at the screen, she thinks David could be even smaller, Goliath yet bigger.

A 2021 study found that less than 4 percent of philanthropic funding dedicated to fighting climate change went to transportation, despite its representing a third of U.S. greenhouse gas emissions. Meanwhile, state agencies were backed by the abiding, indissoluble Highway Trust Fund. Still, the movement was growing. "Freeway fighting work is bringing in a swath of new people. They are people who live in neighborhoods. They see the possibility of, oh, that freeway could go away or here comes the state DOT saying, don't worry we'll relocate you," Martha says. "The local advocates, even so underfunded, outgunned, outmanned, are slowing down highway projects. They are having an impact. I don't think we're successful yet at *stopping*. We have not changed the system . . . yet. But we are making progress."

This book began with a question: If widening highways doesn't fix traffic, why were we spending billions of dollars to widen highways? Like most things in American life, the answer is ideological. For nearly a century, we have been sold the idea that automobiles made us freer and more prosperous. The automobile affirmed the American commitment to individualism—its very name means autonomous machine. A car gave you the ability to control your own personal destiny. What was more American than that?

Eighty years ago, Norman Bel Geddes captivated people with his vision for the future. "Already the automobile has done great things for people," he wrote in his book *Magic Motorways,* published in 1940. "It has taken man out beyond the small confines of the world in which he used to live. Distant communities have been brought closer together. . . . A free-flowing movement of people and goods across our nation is a requirement of modern living and prosperity." Today, we live in the world Bel Geddes once could only imagine. The future of Futurama is uncannily familiar—the highways stretched across the country, stitching together far-flung towns and splicing through cities; the flyover intersections and glass skyscrapers; the automobile's absolute dominance.

Bel Geddes made this world up—he sketched it in charcoal and pitched it to a car company—and then we built it. Today, we need a new Futurama—a vision compelling enough to counter the unsustainable one being sold by car companies and oil executives. "One of the best ways to make a solution understandable to everybody is to make it visual, to dramatize it," Bel Geddes wrote. This is what removing a highway does: It makes visible a different future.

Republicans do not have a monopoly on the belief that highways are essential to economic development and that expanding highways will fix traffic. A dozen states—even those governed by Democratic leaders who have committed to ambitious climate goals—are currently moving forward with massive highway expansions. There are plans on the books to widen highways in New York City, Chicago, San Francisco, and Portland, some of the most liberal cities in the country. Instead of widening highways to fix congestion, states should expand transit service in urban and suburban areas, adding new routes and increasing frequency. Most states already have the flexibility to reallocate money to transit projects, but the federal government should make its preference explicit: In the next long-term surface transportation reauthorization, Congress should fund transit and highways equally—a fifty-fifty split.

Or better yet, Congress should abolish the Highway Trust Fund altogether, funneling gas tax revenue into the general fund and allocating transportation spending according to need and priority, not formula. Fifty years ago, when the Highway Trust Fund was originally set to expire, this was seen as a reasonable idea. "If annual appropriations from the general fund and full, flexible control by the Congress are good enough for national defense, for housing, for education, why are they not good enough for highways?" asked Gaylord

Nelson, a Democratic senator from Wisconsin, in 1972. The following year, when Senator Edmund Muskie argued for opening up the Highway Trust Fund to transit, he called the trust fund "a transportation financing system designed for a time long past." Indeed, the Highway Trust Fund was created for an era when our highest transportation priority was national defense.

Today, our highest transportation priority should be getting people where they are going while reducing greenhouse gas emissions. If we're serious about addressing climate change, we should incentivize electric vehicle production *and* make it so that a majority of Americans can live within walking distance of frequent transit service. In most American cities, transit will remain unreliable and inefficient unless we fund it as if we want people to ride it.

Until Congress acts, change can happen at the state level. State departments of transportation do what their state governments tell them to do. In 2019, the Colorado legislature passed House Bill 1261, which committed the state to reducing greenhouse gas emissions by 90 percent by 2050. After Governor Jared Polis directed all state agencies to make a plan to meet that goal, Colorado's transportation commission approved a rule requiring all new transportation projects to show how they would reduce greenhouse gas emissions, or risk losing funding. Shortly after, the Colorado Department of Transportation announced it was no longer considering adding four lanes to Interstate 25 through the heart of Denver.

And the case for highway removal is gaining momentum. Just east of Rochester, in Syracuse, the New York State Department of Transportation approved a $2.3 billion project to remove an elevated stretch of Interstate 81 east of downtown and replace it with a street grid. "You just have to step back sometimes and say, what were they thinking?" said Governor Kathy Hochul when she announced the funding for the teardown. In 2022, the federal government gave Detroit—Motor City—a $105 million grant to remove I-375, a mile-long sunken highway that traversed the city's downtown, and replace it with a street-level boulevard. When the highway was built in

1964, it displaced thousands of people in Black Bottom, the historic heart of Detroit's Black community. "We don't talk about these harms for the purpose of wallowing. We talk about them because we see a way to fix them," Pete Buttigieg told reporters in Detroit when he announced the grant. (The secretary was in town for the Detroit auto show.) In California, when I-980 began construction through Oakland, public opposition was so fierce that it was abandoned for a decade before the route was finally completed in 1985. In the spring of 2022, the California Department of Transportation began soliciting proposals to remove the two-mile highway.

In the spring of 2023, Congress for the New Urbanism published its biennial *Freeways Without Futures* report, marking fifteen years since the nonprofit began profiling the worst highways in America and highlighting campaigns to remove them. For the first time, the report coincided with an acknowledgment from the federal government of the "inequitable and harmful impacts of urban highway construction." Austin's I-35 was included on the list, its third consecutive appearance.

We should tear down urban highways. But *how* we tear them down also matters. Most urban interstates were designed and built with almost no community input. They should not be removed using the same top-down process. In Washington, D.C., after the District Department of Transportation decided to remove an aging highway bridge that spanned the Anacostia River, a city planner wondered, what about building a park on the highway's abandoned piers? The 11th Street Bridge Park would connect the wealthy, predominantly white neighborhood of Capitol Hill with historic Anacostia, which was majority Black and low income. People in Anacostia worried about all the wealth that would come across the bridge park; they didn't want to be displaced, nor did they want their neighborhood to become unrecognizable to them. So the nonprofit overseeing the 11th Street Bridge Park invested roughly half of its $177 million budget on affordable housing, job training, and community empowerment before the park broke ground. Infrastructure can be conceived not simply as a physical asset but as a community and social one, too.

"The dirty little secret of transportation is that the only thing we count is the speed of vehicles," Beth says. "If you look at most DOTs, they will count that one thing three or four different ways. Time savings, fuel savings. It's all the same measure: How fast are the cars going? The faster, the better." Transportation departments could measure different things, like access to jobs and services. The whole point of transportation, of course, is to arrive.

The first time I talked to Beth, I asked her what motivated her to stick with this work, given that we keep making the same mistakes over and over. "You've got to fight because the opportunities come out of nowhere," she said. "The way change happens is people like me bang our heads up against the wall repeatedly. And we hope that the crack in the wall surfaces before our heads crack. My head might crack before I get the change, but I will set the stage for somebody who will finally push it through. But I'm going to do it because I'm right. Because what we're doing isn't working. And because someday people will wake up and say, 'Why am I spending all this money on something that doesn't give me one of the outcomes I'm seeking?' When that day comes, I will show up and say, let's do something different."

Despite its seeming permanence, the built environment can change quickly—much more quickly than we might imagine.

On a gorgeous spring morning in April 2023—almost exactly three years after the Texas Transportation Commission voted to fund the I-35 expansion—I drive south on the interstate through the center of Austin. Bluebonnets cover the grassy slopes on either side of the highway. Just north of Town Lake, on the eastern edge of I-35, I pull over at 1103 Clermont Avenue, the community land trust property owned by Guadalupe Neighborhood Development Corporation.

Behind a chain-link fence, past piles of construction debris, two houses fill the lot. In front, there is a two-story blue-and-white-sided house shaded by a towering box elder, the tree's serrated green leaves bright and translucent in the spring sunshine. Tucked fifty feet behind the front house sits a small, one-story A-frame house, the exterior sage green with white trim. The interiors still needed to be finished and inspected, Mark Rogers said, but he expected that tenants would move in within a few months, sometime in the summer. A few days earlier, he'd gotten yet another letter from TxDOT, the third or fourth with exactly the same message. "You are probably aware that the Texas Department of Transportation will make improvements to IH-35," it read. "Plans have progressed to the point that we wish to advise you that a portion of your property located at 1103 Clermont Avenue will be needed as right of way for the proposed improvement of the highway. We have attached for your information a booklet entitled 'State Purchase of Right of Way.'" It was the same booklet Modesti Cooper had received four years earlier.

After I leave 1103 Clermont, I drive north along the frontage road, which undulates alongside the highway, passing the little blue house on the hill at Ninth Street that the Lopez family had fought so hard to retain. Just past Martin Luther King Jr. Boulevard, the highway rises over the frontage road, a tidal wave cresting over the city. I drive past the low-slung limestone building that contains Escuelita del Alma. When I get home, I call the school and Jaime Cano answers the phone. He is more optimistic than he was in January. Dina had hired a realtor, who took them to see several properties that might be able to house the school. "They all have their own challenges," Jaime says. Every property would need extensive retrofitting. Two of the buildings were close enough to I-35 that Jaime worried about how construction would affect parents' ability to access the school. "But we've made progress in terms of finding some options. It's not as bleak as it seemed before," he says. "But it's still going to be very difficult to afford."

They hadn't heard anything from TxDOT, Jaime says. Dina, who is sitting next to Jaime, interrupts—actually, a consultant hired by TxDOT had contacted them simply to say she didn't have any additional information on the project's timeline. TxDOT didn't want the school to move before the project had been formally approved, Dina says. Indeed, if they moved too early, they'd forgo any relocation assistance. But Jaime didn't feel as though they could afford to wait and see; hundreds of families relied on their care. "I just feel that if something too good to pass up comes our way, that we're just going to end up forgetting what TxDOT has recommended," he says. In the meantime, Dina had shelved her retirement plans. She wanted to get the school settled in its new home. Then, maybe, she could think about leaving.

Part of Jaime's newfound optimism appeared to be acceptance. The project felt inevitable. "We're going to have to move, so let's just move forward," he says. But still, even as they toured properties, he held out hope that the project would be delayed—"that we don't have to disrupt our community so soon." He was trying to accept their current limbo while preparing for all possible outcomes. "If they say it's not going forward, we're super prepared for that. We're extremely prepared for that option," he says, laughing. "But it just feels like it's going to happen one way or another. TxDOT is going to bulldoze it literally and figuratively."

In August, TxDOT published the final environmental impact statement for the I-35 Capital Express Central project, along with a record of decision. The expansion was approved. Property acquisition could begin and soon after, construction. The day the decision was released, a forty-five-day streak of triple-digit temperatures in Austin finally broke. The summer was on track to be Texas's hottest on record, as a heat dome gripped the state and drought consumed its rivers. TxDOT claimed adding capacity to I-35 would increase greenhouse gas emissions only fractionally. But according to the Rocky Mountain Institute, the effects would be enormous. Widening the highway would generate roughly the annual greenhouse gas emissions of a small coal-fired power plant.

TxDOT estimated the project would take a decade to complete. How hot would it be then?

Back in May, when I caught up with Elda Reyes a month after TxDOT and FHWA signed the voluntary resolution agreement, she hadn't heard that the pause had been lifted. "So it's approved then?" she asked me. Yes, I told her. It's approved. But it was also delayed. According to the resolution, people living in the footprint of the expansion could remain in their homes until construction began—which had been delayed by the transportation commission until at least 2027. Elda and Jesus likely had longer, at least six years. Elda sighed. "I had hope they wouldn't pass it," she said. Over the past two years, she'd held out hope that the federal government wouldn't allow the project to proceed. As the months dragged on, she'd convinced herself that the project might never happen—that somehow it would just go away. Elda and Jesus had finally put a new roof on their house and started other long-delayed repairs. But she was relieved to learn that they could remain for a few more years. She would stay in her home for as long as she could, she told me. It was good that they had more time. "More time to take in that it's happening. That they will widen the highway and we won't be able to live here anymore," she said. It was hard to imagine. "One gets used to the place that they live," she said. She and Jesus were close to the center of Houston, close to their jobs and children. "Our lives will completely change."

Modesti, meanwhile, didn't want anything more to do with the highway. She just wanted to enjoy her home, for as long as she could remain, and do what she could to build up the Fifth Ward community.

On a hot Saturday in early May, the gymnasium at Finnigan Park fills with people. Many are elders from the Fifth Ward, dressed in their churchgoing finest—wide-brimmed hats with big bows, pressed linen pantsuits. Young families arrive with babies in strollers, toddlers on hips. Outside, in the shade cast by the canopy of a gnarled oak tree, three kids assemble oversize blue blocks into a Jenga tower. Despite the years of opposition—despite the federal government's intervention—many people in the Fifth Ward still haven't heard about the North Houston Highway Improvement Project. The voluntary resolution agreement required TxDOT to at least consider changes to the project, and activists wanted to make sure that TxDOT complied with its commitments. Ally Smither, the opera singer turned freeway fighter, had organized this community education fair to inform people about their rights and when construction would begin in their neighborhoods—TxDOT had recently announced the project would take nearly two decades to complete.

Outside the gymnasium, the lush green fields of Finnigan Park are quiet, the only sound the thrumming of cicadas. Sixty years ago, Curley Guidry taught O'Nari Guidry how to drive here, back when the park was full of teenagers lingering during long summer evenings. Today, Curley arrives just after 1 P.M., looking dapper in khaki dress shorts, a black-and-tan striped button-down, and crimson leather driving loafers. Ally and Molly greet him at the door. "We're the people trying to stop the expansion," Ally says, walking him over to a giant foam board set up on an easel, one of a dozen boards arranged around the perimeter of the basketball court. "The idea for today is people don't know what's going on, they don't know when construction is starting, they don't know what their rights are," she says. "We're giving everyone all that information. We're also getting to know people. Because we still have a few years before they break ground in this area, and we think we can still stop them. Or if we can't stop them, we can make sure that we fight for better." She asks Curley where he lives. "Are they taking your house?" she says.

"No, I'm not in the plan," Curley says. "But they already wiped out my neighborhood."

They stand in front of a foam board with a map of segment three alongside a detailed construction timeline. Construction in the Fifth Ward wouldn't begin until 2027, Ally says. Had Curley heard about the voluntary resolution agreement between TxDOT and the federal government? she asks.

"I ain't hearing nothing about no agreements," he says. "They already took people out of Kelly Village."

"Basically, they're making TxDOT do more to help people impacted by the project," Ally says.

"That should be the backbone, you understand what I'm saying? Don't nobody want to move," Curley says.

Ally's phone rings and she excuses herself and rushes outside— her husband, Austin, is driving a van to transport elders to and from their homes—and Curley and I circulate slowly through the gymnasium, pausing at each station showcasing a different segment of the massive expansion.

In a corner farthest from the door, Michael Moritz stands behind a folding table cluttered with clipboards and ballpoint pens. He's collecting signatures for a ballot initiative called Fair for Houston, which would reform the Houston-Galveston Area Council, the metropolitan planning organization that had voted to fund the North Houston Highway Improvement Project back in 2019. "One of the reasons why projects like the I-45 expansion keep happening in our city is because of the Houston-Galveston Area Council," Michael says. A Black man in his sixties wearing a Dallas Cowboys beanie and white cargo pants listens, rapt. "As a group, their responsibility is to allocate federal infrastructure money for transportation, flood mitigation, workforce development," he says. "Houston and Harris County are vastly underrepresented. That manifests itself negatively. We're left out of funding. Or we're forced to endure freeway expansions that benefit rural and suburban counties."

Houston and Harris County made up nearly 60 percent of the population served by the Houston-Galveston Area Council, and yet city and county representatives controlled only 11 percent of the council's seats. Thirty-seven members sat on the council's board—only two of whom were elected by Houston voters. In 2022, when the council allocated $488 million in federal flood mitigation money, just 2 percent had been allocated to Houston. And the Houston-Galveston Area Council had consistently endorsed the I-45 expansion—despite representatives from Houston and Harris County repeatedly voting against it.

Fair for Houston wanted to change that. Since January, Molly, Michael, and nearly a hundred other volunteers had canvased Houston, collecting thousands of signatures for a ballot measure that would force the regional body to adopt proportional voting, giving Houston and Harris County more seats on the council. The campaign represented a fundamental shift for Stop TxDOT I-45, the grassroots group founded with a single mandate: oppose. Now, the group sought affirmative change. Next time, it would be different.

Just after 2 P.M., Kendra London calls people to sit in rows of folding chairs facing a projection screen set up under a basketball hoop. Curley and I sit down a few rows from the front. A 2018 film called *Ramps to Nowhere* flickers on the screen. The documentary tells the story of a citywide freeway revolt that gripped Seattle in the late 1960s, after the Washington Department of Highways proposed crisscrossing the city with a dense highway network. Plans showed a north-south highway called the R. H. Thomson Expressway cutting through a predominantly Black neighborhood called Central District. A coalition of white and Black activists—including the nascent Black Panther Party—demanded that the city stop the demolition of Black neighborhoods. Thousands of people crowded city hall during several public meetings to register their opposition to the freeway plan. The outcry worked. Although the state department of transportation had already begun building on-ramps connecting the existing Highway 520 with the proposed R. H. Thomson Expressway, in 1970, the Seattle City Council voted to remove the expressway from the city's comprehensive plan, killing the project. The R. H. Thomson Expressway was never built. For decades, those ramps to nowhere remained, beams rising from the bay—a visual reminder that highways could be stopped, even after they'd been started.

After the credits roll and Ally flips the lights back on, an older man sitting near the front raises his hand. "I'd like to say something," he says. "What is the purpose of us all trying to stop I-45 if actually it's a done deal?"

Because it's not over until it's over, Kendra says. Her grandmother, Ella Morris, sits behind a folding table on one side of the gymnasium, eating barbecue out of a Styrofoam container. In the 1960s, highways simply happened to people like Ella and Curley, who woke up one day to find houses gone, their neighborhoods changed beyond recognition. And Kendra's grandmother was one of the lucky ones. Unlike O'Nari, she got to stay. Kendra wanted to believe that things were different today. "If we have a chance at stopping it, why not? Why not work together to stop it? It's our home," Kendra says. The man in the Dallas Cowboys beanie raises his hand. His voice is inaudible until Ally passes him the microphone. "Thank you for being here today," he says, his voice reverberating against the hardwood floor. "If we fight this together, we can win this together."

EPILOGUE

In 1999, to accommodate a booming city, Austin's Robert Mueller Municipal Airport was closed and planes were rerouted to the brand-new Austin-Bergstrom International Airport fifteen miles to the south, leaving the City of Austin with 711 acres of land in the middle of the city, a rare opportunity to do something of scale. In 2004, the city approved a redevelopment plan to transform the former airport into a mixed-income, mixed-use neighborhood, with housing and offices, restaurants and retail, parks and playgrounds, developed as a "sustainable, transit-oriented community" with a "broad range of homes." According to the plan, 25 percent of homes built would be sold at affordable prices to families making below the area's median income.

As a journalist, I earned well below this threshold. I spent weeks documenting my finances to submit to the nonprofit that managed the affordable homes program. After I was approved, I called a representative for CalAtlantic Homes, one of two affordable home builders in the neighborhood. There was one town house still available in my income bracket, did I want to come see it today?

Now I live in a place that used to be an airport. I can live in this place only because a generation ago someone decided that a mixed-income, walkable community was a goal worth pursuing. Because someone looked around the city and realized that this goal wasn't going to happen on its own—that if we wanted a street full of elementary school teachers and hospital nurses and social workers, as my street has, then we'd have to engineer it. I don't own the land and can't earn a profit on my house; it is essentially a community land trust, limited in how much it can appreciate. But my whole life changed when I bought my house. Suddenly I could imagine a future for myself in Austin. No one was going to raise the rent and make me move.

Mueller was still being built when I moved in 2020. New town houses sprouted to the east of me, one after another. It was the beginning of the pandemic, and as I walked around the neighborhood, over weeks and then months, I watched the neighborhood emerge. By 2021, new streets cut across what was once a field of grass and bramble; a few months later, the frames of houses began to grow from cement foundations. Gradually, I met my new neighbors. One guy in his early forties simply called at me from his stoop, "Hey, who are you?" We were all new to the neighborhood; there was no etiquette. He introduced me to other neighbors, and soon we were dragging lawn chairs and beers into the grassy median that separated the two sides of the street. When Texas froze over for one harrowing week, we texted to remind each other where our water shutoff valves were, to check our attics for snow, to make sure everyone had enough food. After four cooped-up days, someone built a tiny fire outside and we dragged our lawn chairs into the snow-covered street to drink wine out of cold mugs.

In the summer of 2021, my parents retired, sold the house I grew up in, and moved into one of the brand-new houses that had sprouted a few blocks from my own. My sister had her third daughter during the pandemic, and my parents were fed up with the distance from their grandchildren. Gradually, they settled into retirement and back into life in Austin, which by then was a completely different place from the one they had left three decades earlier.

A year after they moved, my dad was walking around the neighborhood, as he did almost every morning, when he fell and hit his head. After the fall, he became confused, disoriented. The fall revealed other, more serious problems. His doctor told him to stop driving; it was no longer safe, each trip too large a withdrawal from his dwindling risk bank. My dad had always been the family driver. He drove me and my sister to high school, to basketball games, and out to movies with our friends. After I moved to Tucson, whenever I flew home, my dad picked me up at LAX and dropped me back off, days later. One Christmas, he named his taxi service Harry's Happy Holidays, or 3H. The name stuck. After my parents moved to Austin, my dad resumed his airport shuttle service. Even at 6:00 A.M., he'd roll up at my doorstep in the predawn darkness, delighted to help. "3H at your service," he'd say. But after the fall, I watched as my mom became the sole keeper of their mobility. Just as I had when I moved to Austin, suddenly she was spending hours of her day isolated in the car, stretches of time that she might have spent doing all the things she imagined for her own retirement.

To lose your ability to drive in a car-centric city is to forsake all sense of yourself as an independent and autonomous person. The loss of control made my dad an anxious passenger, overbearing and interfering. The highways terrified him. The cars drove too fast, changed lanes unpredictably. My mom was enduringly patient, but sometimes she got angry—with him and the highways but mostly with the circumstances that landed her in the driver's seat. I was angry too—furious that there was no other option but for her to drive.

In early 2022, orange construction barrels went up at the end of my street. Men and women in hard hats showed up and rerouted traffic a block east. The sidewalk expanded into a wide slab surrounded by a white metal guardrail. And then, one day, a sign appeared: PROJECT CONNECT COMING SOON! It was a bus stop, part of a new MetroRapid line that would stretch from northeast Austin to downtown. As Project Connect plans progressed, the city published renderings showing light-rail cars moving throughout the city, new stations sprouting on familiar streets. The fully realized reality was years away, scaled back and further delayed by inflation, but the renderings—a car-free "transit mall" on Guadalupe Street, a light-rail vehicle emerging from a tunnel downtown, a bus on a bridge spanning Town Lake—offered a glimpse of a different future. I wanted to live there, in *that* city. I wanted my parents to live there, too.

Infrastructure seems inevitable. It appears permanent. But twenty years ago, someone walked around an empty airport and envisioned the life I now inhabit. I am captivated by the idea of highway removal—the idea of replacing the violent, dangerous structures that contour our cities with parks and promenades and places to live—because it makes so much else possible. If the isolation of the pandemic taught us anything, it is how desperately we need to be around each other—even people we don't know, or perhaps *especially* people we don't know, the strangers who thread our lives with possibility.

Tearing down a highway is just the beginning.

ACKNOWLEDGMENTS

This book would not exist without the dozens of generous, brilliant people who shared their time and expertise with me. Foremost among them are Jay Blazek Crossley and Heyden Black Walker, who helped shape this book early on. Thank you, Jay, for your boundless wonkiness, and Heyden—and the entire Black family—for believing I-35 could be better. Mostly, highways are inert objects—a reporting challenge—but Adam Greenfield ignited movement. Without Adam, the fight would be so much less interesting. Thanks to Beth Osborne, who Steve Davis called "terrifyingly smart," and Ben Crowther—both of their insights are woven throughout this book. I met Patrick Kennedy standing under I-345 in early 2021 and was captivated by his vision. Three years later, I still am.

I am overwhelmed by the generosity of those who opened their homes and lives to me, and I am enormously grateful for their trust. Thank you to Modesti Cooper, O'Nari Guidry, Curley Guidry, Ella Morris, Kendra London, Jayne McCullough, James Caldwell, Jasmine Gaston, and the many Fifth Ward residents, past and present, who spoke with me about their cherished neighborhood. To Dina Flores and Jaime Cano: Thank you for inviting me into your school and trusting me to observe all the little lives you caretake. Your tenacity and dedication are an inspiration. I first interviewed Mark Rogers in 2018 and he remains one of my favorite people to talk to. Thank you to Mary Helen Lopez for sharing your family's story with me.

Molly Cook is a force of nature. Thank you to Molly, Mark, Judy, and Chloe Cook, for inviting me into your home and family. Thank you to the Stop TxDOT I-45 team for letting me be a fly on the wall for three years, most especially Susan Graham, Michael Moritz, and Ally Smither. Special thanks to Harrison Humphreys at Air Alliance Houston.

Jerry Hawkins taught me about the hidden history of Dallas. Donald Payton helped me find it. Thanks to Maggie Parker for pointing me to the Forest Theater—and to Elizabeth Wattley, LaSheryl Walker, and Lockridge and Lakeem Wilson for sharing its history. Thank you to Mo Bur and Ceason Clemens in the TxDOT Dallas office for your willingness to engage. A heartfelt thank you to the former TxDOT executives who spoke with me on background and trusted me enough to be frank. In Rochester, I am indebted to Shawn Dunwoody and Suzanne Meyer.

I spent countless hours in library archives, and I am grateful to all those who help caretake our collective history. Thanks, particularly, to the staff at the Dallas Public Library, the Houston Public Library, the Austin History Center, the Rochester Public Library, and the National Archives, in Abilene, Kansas, and College Park, Maryland.

To write this book, I relied on the work of other journalists, notably Dug Begley at the *Houston Chronicle* and Matt Goodman at *D Magazine*. Thanks to Tristan Ahtone for editing my story in *The Texas Observer*, which allowed me to explore this topic, and for the unflappable support as I navigated writing the proposal that became this book. Jessica Goudeau and Roxanna Asgarian read drafts in the book's infancy and believed in its potential.

It turns out that the best place to write about highways is nowhere near them. Thank you to the Carey Institute for Global Good for hosting me as a Logan Nonfiction Fellow, and to my fellow journalists for the camaraderie and inspiration. Rafil Kroll-Zaidi provided astute feedback on early outlines. I wrote a fourth of this book looking out over the Tomales Bay from the west cabin at the Mesa Refuge in Point Reyes, California. Thank you to Peter Barnes for creating and caretaking such a beautiful, productive space for writers and activists. Thank you to Katherine and Eric at the Tasajillo Residency for hosting me and my dog on a gorgeous ranch outside of Kyle—a straight shot down I-35 from my home in Austin. I was so lucky to benefit from such wild, open space while writing about the very thing that threatens it.

Ten years ago, I sent a cold query to Mackenzie Brady Watson. Since then, she has shepherded two books into being. I'm still awestruck with gratitude to have Mackenzie in my corner. Thank you, Mackenzie, for believing in me and my writing—and for your patience and advice during the many years it took for this book to come into being. Our relationship is one of the greatest gifts of my career. I am honored to be a part of the Stuart Krichevsky Literary Agency, which has provided this project with incalculable support. Thank you to my amazing editor, Aubrey Martinson, for sharing my vision and for your steadfast commitment to this project and my process, even when I doubted. The book is so much better for your editing. Thank you to Crown for giving this book such a good home. Hilary McClellen provided excellent fact-checking.

Thank you to Nora Fallon and Joshua Albert, whose friendship carried me while I wrote this book. Jacob Barnes and Amber Jekot reminded me to go outside. Amy Wilde read a draft and her enthusiasm got me across the finish line.

This book is for my family. Katie and Tyler Grooms got me to Austin and have supported my life and career in countless ways, large and small. Ellie, Bryn, and Madeline Grooms kept me here and made this city feel like home. I started saying that I wanted to be a writer when I was in the third grade, and my parents have never wavered in their support of that dream. Midge Kimble taught me how to talk to absolutely anyone—how to see the good and believe in better. Thank you for your encouragement and help when things were hard. To Jeff Kimble, better known as 3H: Thank you for teaching me how to marvel at the world and to be relentlessly curious—and for driving me everywhere I ever needed to go.

NOTES

Chapter 1: Home

5 *"State Purchase of Right of Way"*: "State Purchase of Right of Way" (brochure 15.500), Right of Way Division, Texas Department of Transportation, Oct. 2012, 2.

Chapter 2: Distance

7 *One morning in April:* Texas Transportation Commission public meeting, April 30, 2020, txdot.new.swagit.com/videos/168586.

7 *eighty thousand miles of roadway:* "Roadway Inventory Annual Reports 2021," Texas Department of Transportation, 1.

7 *more than any other state:* "Estimated Lane Miles by Functional System," Federal Highway Administration, Oct. 2009, www.fhwa.dot.gov/policyinformation/ statistics/2008/hm60.cfm.

7 *a banker from San Antonio:* "J. Bruce Bugg, Jr., Chairman," Texas Department of Transportation, www.txdot.gov/about/leadership/texas-transportation -commission/j-bruce-bugg-jr-chairman.html.

7 *city's population approached 300,000:* City of Austin Population History, City of Austin, www.austintexas.gov/sites/default/files/files/Planning/Demographics/ population_history_pub.pdf.

7 *double-deck interstate:* Jimmy Moss, "Why Was I-35 Designed as a Double-Decker Through Central Austin?," KUT, Jan. 19, 2018, www.kut.org/transportation/2018 -01-19/why-was-i-35-designed-as-a-double-decker-through-central-austin.

7 *the Austin metropolitan region:* Dinah Chukwu, "Austin Named Fastest Growing Major Metro in the US," KVUE, May 4, 2021, www.kvue.com/article/money/ economy/boomtown-2040/austin-population-growth-census-data/269-c1e8725e -3489-4445-9bb5-fc340887cc43.

7 *2 million people:* U.S. Census, Austin–Round Rock–Georgetown, Texas Metro Area, censusreporter.org/profiles/31000US12420-austin-round-rock-georgetown-tx -metro-area/.

8 *most congested stretch of highway:* "100 Most Congested Roadways in Texas," Texas A&M Transportation Institute Mobility Division (2020), mobility.tamu.edu/texas -most-congested-roadways/.

8 *fourth in the nation:* "National List of Major Freight Highway Bottlenecks and Congested Corridors," Freight Mobility Trends: Truck Hours of Delay, Federal Highway Administration (2019), ops.fhwa.dot.gov/freight/freight_analysis/ mobility_trends/national_list_2019.htm.

8 *Most of the roads:* Gary Scharrer, *Connecting Texas: True Tales of the People Who Built Our Highways and Bridges* (Austin: Greenleaf Book Group, 2020), 3.

9 *"I do not believe mass transit":* Griffin Smith, Jr., "The Highway Establishment and How It Grew and Grew and Grew," *Texas Monthly,* April 1974.

9 *"A guy in a wheelchair":* Ken Herman, "Did Greg Abbott Really Wheel near Traffic for Campaign Ad?," *Austin American-Statesman,* Oct. 3, 2014, www.statesman.com/ story/news/2014/10/03/herman-did-greg-abbott-really-wheel-near-traffic-for -campaign-ad/10036128007/.

9 *turn dirt to build roads:* "Texas Transportation Commission Chairman J. Bruce Bugg, Jr., Interviews Texas Governor Greg Abbott," Texas Transportation e-Forum, Feb. 9, 2021, www.youtube.com/watch?v=XNWt9lLT8Fk.

9 *Between 2015 and 2023:* Texas Transportation Commission public meeting, March 29, 2023, txdot.new.swagit.com/videos/223077.

9 *the deadliest road in the city:* Maeve Ashbrook, "I-35 Is the Deadliest Road in Austin, New Study Reveals," KVUE, July 6, 2021, www.kvue.com/article/traffic/i-35 -named-deadliest-road-in-austin/269-683546fb-764b-4373-bf8a-e7a32ca1e131.

10 *Arroyo Seco Parkway:* "California SP Arroyo Seco Parkway Historic District," National Register of Historic Places, National Park Service, catalog.archives.gov/id/ 123858927.

12 *Herbert Hoover . . . declared:* Sarah A. Seo, *Policing the Open Road: How Cars Transformed American Freedom* (Cambridge, Mass.: Harvard University Press, 2019), 39.

13 *Transportation is the leading source:* "Fast Facts on Transportation Greenhouse Gas Emissions," U.S. Environmental Protection Agency, www.epa.gov/greenvehicles/ fast-facts-transportation-greenhouse-gas-emissions.

13 *In Austin, the average driver:* Nadja Popovich and Denise Lu, "The Most Detailed Map of Auto Emissions in America," *New York Times,* Oct. 10, 2019, www.nytimes .com/interactive/2019/10/10/climate/driving-emissions-map.html.

13 *on-road emissions:* "Statewide On-Road Greenhouse Gas Emissions Analysis and Climate Change Assessment," Texas Department of Transportation, Oct. 2018, 7.

13 *Highway widenings in Texas:* "State-Level Results: Emissions Analysis of BIL Investment Scenarios," Georgetown Climate Center and Rocky Mountain Institute, March 2023.

13 *the average family in Houston:* Center for Neighborhood Technology Housing and Transportation (H+T®) Affordability Index: Houston, New York City, htaindex.cnt .org.

13 *the longer average commutes:* Raj Chetty and Nathaniel Hendren, "The Impacts of Neighborhoods on Intergenerational Mobility," Harvard University, May 2015, scholar.harvard.edu/files/hendren/files/nbhds_paper.pdf.

14 *the human cost:* "Traffic Safety Facts Texas 2017–2021," National Highway Traffic Safety Administration, 4.

14 *a panel convened:* "Traffic-Related Air Pollution: A Critical Review of the Literature on Emissions, Exposure, and Health Effects: Executive Summary," Health Effects Institute, Jan. 2010, 3.

14 *fifty-three thousand early deaths:* "Air Pollution and Early Deaths in the United States," *Atmospheric Environment* (Nov. 2013): 198–208, www.sciencedirect.com/ journal/atmospheric-environment.

14 *travel-time budget:* Yacov Zahavi, "Traveltime Budgets and Mobility in Urban Areas," U.S. Department of Transportation, May 1974.

14 *"Higher speeds enable life":* Tom Vanderbilt, *Traffic: Why We Drive the Way We Do (and What It Says About Us)* (New York: Alfred A. Knopf, 2008), 274.

15 *In 1962, an economist:* Anthony Downs, "The Law of Peak-Hour Expressway Congestion," *Traffic Quarterly* 16, no. 3 (1962): 393–409.

15 *In the 1960s, he worked for:* Matt Schedel, "Anthony Downs, Who Viewed Politics

and Traffic Through the Lens of Economics, Dies at 90," *Washington Post,* Oct. 27, 2021, www.washingtonpost.com/local/obituaries/anthony-downs-dead/2021/10/27/ebff5d5a-3679-11ec-8be3-e14aaacfa8ac_story.html.

16 *the hundred largest urbanized areas:* "The Congestion Con," Transportation for America, March 2020, 4.

16 *What is now Interstate 35:* Michael Barnes, "Recalling Austin's Ample East Avenue," *Austin American-Statesman,* May 31, 2014.

17 *"in the vicinity of East Avenue":* "Council Hears Proposals for Super-Highway," *Austin American,* July 13, 1945.

17 *"after Greer frankly told them":* William J. Weeg, "City Council Agrees to Buy Super Highway Right-of-Way," *Austin Statesman,* Sept. 26, 1946.

17 *Austin needed a "super highway":* Morris Midkiff, "Austin Must Begin Now on Super Highway," *Austin Statesman,* Jan. 30, 1944.

17 *Austin's first comprehensive plan:* Koch & Fowler, "A City Plan for Austin, Texas," Jan. 1928.

17 *the new Interregional Highway:* "Super-Duper Begun," news clipping, April 19, 1950, Austin History Center.

17 *In 1958, the Texas Highway Department:* Analysis by author of Texas Department of Transportation Right of Way Division records, Texas Digital Archive.

17 *half a million homes:* "Beyond Traffic 2045," U.S. Department of Transportation (2017), 95, www.transportation.gov/sites/dot.gov/files/docs/BeyondTraffic_tagged_508_final.pdf.

18 *"The effects of the project":* "Community Impacts Assessment, North Houston Highway Improvement Project," Texas Department of Transportation, August 2020, sec. 5.9.2, pp. 5–218.

18 *freeway fighters erased highway lines:* Mark H. Rose and Raymond A. Mohl, *Interstate: Highway Politics and Policy Since 1939,* 3rd ed. (Knoxville: University of Tennessee Press, 2012), 113–33.

18 *In Washington, D.C.:* Chris Myers Asch and George Derek Musgrove, *Chocolate City: A History of Race and Democracy in the Nation's Capital* (Chapel Hill: University of North Carolina Press, 2017), 362–66.

19 *In Baltimore:* Raymond A. Mohl, "The Interstates and the Cities: Highways, Housing, and the Freeway Revolt" (Poverty & Race Research Action Council, 2002), 26.

19 *In New York City:* Dwight Garner, "When David Fought Goliath in Washington Square Park," *New York Times,* Aug. 4, 2009, www.nytimes.com/2009/08/05/books/05garner.html.

19 *city leaders approved a plan:* "Austin Historic Freeway Planning Maps," TexasFreeways.com, www.texasfreeway.com/austin/austin-historic-freeway-planning-maps/.

19 *People were horrified:* Record Group 30: Records of the Bureau of Public Roads, 1892–1972, boxes 1421–24, National Archives at College Park, College Park, Maryland.

20 *People crowded city hall:* "Proposed Freeway Debated," *Austin Statesman,* Feb. 17, 1967.

Chapter 3: Future

21 *When automobiles first trundled:* Kenneth T. Jackson, *Crabgrass Frontier: The Suburbanization of the United States* (New York: Oxford University Press, 1985), 157–58.

21 *the year the Model T:* "State Motor Vehicle Registrations, by Years, 1900–1995," Federal Highway Administration, www.fhwa.dot.gov/ohim/summary95/mv200.pdf.

21 *Shell Oil hired:* B. Alexandra Szerlip, *The Man Who Designed the Future: Norman Bel Geddes and the Invention of Twentieth-Century America* (New York: Melville House, 2017), 215–30.

22 *eventually spending $6.7 million:* Ibid., 246.

22 *New York World's Fair:* Lawrence Speck, "Futurama," in *Norman Bel Geddes Designs America,* ed. Donald Albrecht (New York: Harry N. Abrams, 2012), 289–97.

24 *The new American community:* Frank Lloyd Wright, "Broadacre City: A New Community Plan," *Architectural Record* (1935).

24 *American Road Builders' Association:* Jackson, *Crabgrass Frontier,* 248.

25 *"the automobile has brought more progress":* Ibid., 162.

25 *The Triborough had opened:* "200,000 Rush to Use New Bridge by Auto, Bus, Cycle, and on Foot," *New York Times,* July 12, 1936.

25 *Moses took the stage:* "Text of Addresses by Roosevelt, Lehman, and Others at Bridge Ceremony," *New York Times,* July 12, 1936.

25 *"a new freedom from congestion":* "New Span Climax of 23-Year Effort," *New York Times,* July 12, 1936.

25 *Moses opened three expansive:* Robert Caro, *The Power Broker: Robert Moses and the Fall of New York* (New York: Vintage Books, 1975), 515–18.

26 *"special system of direct interregional highways":* Toll Roads and Free Roads, 76th Cong., 1st Sess., House Document No. 272 (1939).

27 *the West Hall:* Rose and Mohl, *Interstate,* 11.

27 *Bel Geddes wrote to Roosevelt:* Speck, "Futurama."

27 *highways since 1916:* Richard F. Weingroff, "Federal Aid Road Act of 1916: Building the Foundation," *Public Roads* 60, no. 1 (1996), highways.dot.gov/public-roads/summer-1996/federal-aid-road-act-1916-building-foundation.

27 *In 1944, Congress passed:* "House Approves $1,500,000,000 Program to Modernize Nation's Roads After War," *New York Times,* Nov. 30, 1944.

27 *Transcontinental Motor Convoy:* "1919 Transcontinental Motor Convoy," Dwight D. Eisenhower Presidential Library, www.eisenhowerlibrary.gov/research/online-documents/1919-transcontinental-motor-convoy.

27 *the German Autobahn:* J. Richard Capka, "Celebrating 50 Years: The Eisenhower Interstate Highway System," remarks before the House Committee on Transportation and Infrastructure, June 27, 2006, www7.transportation.gov/testimony/celebrating-50-years-eisenhower-interstate-highway-system.

27 *Eisenhower spoke to Congress:* "Message to the Congress Regarding Highways, February 22, 1955," White House Office, Office of the Press Secretary to the President, box 4, Eisenhower Presidential Library, www.eisenhowerlibrary.gov/research/online-documents/interstate-highway-system.

28 *In 1956, Eisenhower signed:* National Interstate and Defense Highways Act (1956), National Archives, www.archives.gov/milestone-documents/national-interstate-and-defense-highways-act.

28 *Eisenhower appointed:* "Maj. Gen. John S. Bragdon, 70, of Civil Aeronautics Board Dies," *New York Times,* Jan. 8, 1964.

29 *the president tasked him:* "Letters of Appointment," July 1, 1955, John Stewart Bragdon Papers, 1954–1962, Eisenhower Presidential Library.

29 *wrote to Eisenhower:* "Memorandum for the President," June 17, 1959, John Stewart Bragdon Records, 1949–1961, Eisenhower Presidential Library.

29 *These urban highways:* William S. Odlin, Jr., "Highway Plans at Crossroads as Costs Soar to the Skies," *Transport Topics,* Oct. 12, 1959, John Stewart Bragdon Records, 1949–1961, Eisenhower Presidential Library.

29 *Eisenhower tasked his deputy:* "Highway Review Program Directive of the President," July 2, 1959, Bragdon Records, 1949–1961.

30 *received a report:* "Legislative Intent with Respect to the Location of Interstate Routes in Urban Areas," Bureau of the Budget and the Department of Commerce, Sept.1959, Bragdon Records, 1949–1961.

30 *Tallamy was an engineer:* Caro, *Power Broker,* 11.

31 *a typical day:* "The President's Appointments Wednesday April 6, 1960," Dwight D. Eisenhower: Papers as President (Ann Whitman File), 1953–1962, Eisenhower Presidential Library.

31 *Bragdon recounted:* "Text for Chart Presentation for Interim Report," April 4, 1960, Bragdon Records, 1949–1961.

32 *Eisenhower's response:* "Memorandum for the Record, Meeting in the President's Office—Interim Report on the Interstate Highway Program," April 8, 1960, Dwight D. Eisenhower: Papers as President (Ann Whitman File), 1953–1962.

33 *Ann Whitman:* "Diary, April 6, 1960," ACW Diary, Dwight D. Eisenhower: Papers as President (Ann Whitman File), 1953–1962.

33 *president of Colombia:* "Memorandum of Conversation: Call on the President by President Lleras Camargo," April 6, 1960, Dwight D. Eisenhower: Papers as President (Ann Whitman File), 1953–1962.

34 *"General Bragdon's last major":* "Exit Bragdon: White House Post Vacated as General Moves to CAB," *Engineering News-Record,* May 5, 1960, Bragdon Records, 1949–1961.

Chapter 4: Blight

35 *Houston's Fifth Ward:* "Super Neighborhood 55—Greater Fifth Ward," City of Houston, www.houstontx.gov/superneighborhoods/55.html; Patricia Pando, "In the Nickel, Houston's Fifth Ward," *Houston History* 8, no. 3 (2011).

35 *Creoles of color:* Diana J. Kleiner and Ron Bass, "Frenchtown, Houston," *Handbook of Texas Online,* Jan. 1, 1995, www.tshaonline.org/handbook/entries/frenchtown -houston.

39 *Interstate 10 opened:* "Freeway Section Opens," *Houston Chronicle,* April 18, 1966.

40 *1,220 structures:* Kyle Shelton, *Power Moves: Transportation, Politics, and Development in Houston* (Austin: University of Texas Press, 2017), 61, 81.

40 *remove "slums":* Toll Roads and Free Roads.

41 *national housing emergency:* "Statement by the President on the Veterans' Emergency Housing Program," Feb. 8, 1946, www.presidency.ucsb.edu/documents/ statement-the-president-the-veterans-emergency-housing-program-0.

41 *a U.S. Navy veteran:* Colin Marshall, "Levittown, the Prototypical American Suburb," *Guardian,* April 28, 2015, www.theguardian.com/cities/2015/apr/28/ levittown-america-prototypical-suburb-history-cities.

42 *federally backed mortgages:* Richard Rothstein, *The Color of Law: A Forgotten History of How Our Government Segregated America* (New York: Liveright, 2017), 70–71.

42 *"aid in the efficient assembly":* Richard F. Weingroff, "Designating the Urban Interstates," Highway History: Federal Highway Administration, www.fhwa.dot.gov/ infrastructure/fairbank.cfm.

42 *"some city officials expressed":* Rothstein, *Color of Law,* 127–28.

42 *The HOLC was created:* Ibid., 63–65.

42 *Standardized maps rated:* Robert K. Nelson et al., "Mapping Inequality," in *American Panorama,* ed. Robert K. Nelson and Edward L. Ayers, dsl.richmond.edu/ panorama/redlining/#loc=12/29.746/-95.416&city=houston-tx.

43 *redlining maps with interstate routes:* Clayton Nall and Zachary P. O'Keeffe, "What Did Interstate Highways Do to Urban Neighborhoods?," unpublished, Nov. 21, 2018, www.nallresearch.com/uploads/7/9/1/7/7917910/urbanhighways.pdf.

43 *highway construction demolished:* Eric Avila, *The Folklore of the Freeway: Race and Revolt in the Modernist City* (Minneapolis: University of Minnesota Press, 2014), 20.

43 *more than a million people:* "Beyond Traffic 2045," U.S. Department of Transportation (2017), 95.

Chapter 5: Remove

45 *The first highway:* "Expressway Dedication Set Tonight," *Dallas Morning News,* Aug. 18, 1949; Warren Leslie, "Expressway Opens," *Dallas Morning News,* Aug. 20, 1949.

45 *fifty million vehicles:* "State Motor Vehicle Registrations, by Years, 1900–1995."

45 *the city convened: Report to Dallas,* Citizens Traffic Commission (1955), www .youtube.com/watch?v=leAKXcaLDd4.

46 *North Dallas:* Cynthia Lewis, "Under Asphalt and Concrete: Post-war Urban Redevelopment in Dallas and Its Impact on Black Communities, 1943–1983" (master's thesis, Texas Woman's University, 2019), 1–2.

46 *tore through the center:* "Central Boulevard to Take In Homes of Long-Time Residents," *Dallas Express,* Oct. 5, 1946.

46 *turned its attention south:* "Central Expressway South Dallas Asset," *Dallas Morning News,* March 22, 1951.

46 *predominantly white neighborhood:* Jim Schutze, *The Accommodation: The Politics of Race in an American City* (Dallas: Deep Vellum, 2021), 17.

47 *a grand theater:* "5,000 Jam Opening of New Forest," *Dallas Morning News,* July 30, 1949; "New Forest Features Modern Architectural Innovations," *Dallas Morning News,* July 28, 1949.

47 *workers started pulling up:* "Tracks Come Up on South Central," *Dallas Morning News,* June 23, 1954.

47 *an aerial photograph:* Paula Bosse, "South Central Expressway Under Construction—1955," *Flashback Dallas* (blog), May 14, 2016, flashbackdallas.com/ 2016/05/14/south-central-expressway-under-construction-1955/.

47 *More than thirteen hundred homes:* "The History of the Forest Theater," ForestForward.com/history.

47 *"de luxe movie house for Negroes":* "Forest Theater Has Reopening for Negro Use," *Dallas Morning News,* March 3, 1956.

48 *an elevated highway:* Allen Quinn, "Dallas Seeks Aid for Elevated Artery," *Dallas Morning News,* Sept. 11, 1956.

48 *what had happened:* "Expressway Group Asks Project Report," *Dallas Morning News,* Dec. 10, 1959.

48 *interstate highway network:* "Freeway Gets Green Light," *Dallas Morning News,* Oct. 17, 1964.

48 *Interstate 345:* "State Agency OK's Work on Elevated Road," *Dallas Morning News,* June 13, 1968.

48 *"one of the most imposing structures":* Gene Ormsby, "5-Mile Bridge Nears," *Dallas Morning News,* May 19, 1968.

48 *Hundreds of fatigue cracks:* Amy Smith Barrett, Hyeong Kim, and Karl H. Frank, "Field Test and Finite Element of I-345 Bridge in Dallas," Center for Transportation Research, University of Texas at Austin (2009), library.ctr.utexas.edu/ctr -publications/5-4124-01-2.pdf.

49 *master plan for downtown:* "Downtown Dallas 360—a Pathway to the Future," City of Dallas (2011), 13, www.downtowndallas360.com/wp-content/uploads/2015/ 06/Dallas360_Final-1Introduction.pdf.

49 *hundreds of acres:* Patrick Kennedy, "How Dallas Is Throwing Away $4 Billion," *D Magazine,* Jan. 22, 2013, www.dmagazine.com/publications/d-magazine/2013/ february/how-dallas-is-throwing-away-4-billion-dollar-investment/.

50 *Dallas decidedly was not:* U.S. Census QuickFacts, Dallas City, Texas, population 2000–2010.

50 *Dallas's downtown office buildings:* Connie Gore, "The Rejuvenation of Downtown

Dallas," *D Magazine,* Sept. 15, 2010, www.dmagazine.com/publications/d-ceo/2010/october/the-rejuvenation-of-downtown-dallas/.

50 *"Here's a dirty little secret":* Kennedy, "How Dallas Is Throwing Away $4 Billion."

50 *public meeting about I-345:* Robert Wilonsky, "At TxDOT Hearing, Leaving Comments and Making the Case for Tearing Down Highway Separating Deep Ellum, Downtown," *Dallas Morning News,* Dec. 12, 2012, www.dallasnews.com/news/transportation/2012/12/12/at-txdot-hearing-leaving-comments-and-making-the-case-for-tearing-down-highway-separating-deep-ellum-downtown/.

51 *study called CityMAP:* "Dallas City Center Master Assessment Process," Texas Department of Transportation (2016), www.dallascitymap.com.

51 *Embarcadero Freeway:* Richard F. Weingross, "Essential to the National Interest," *Public Roads* 69, no. 5 (2006), highways.dot.gov/public-roads/marchapril-2006/essential-national-interest.

52 *city leaders proposed:* Vanessa Arredondo, "The Rise and Demise of San Francisco's 'Hideous' Embarcadero Freeway," *San Francisco Chronicle,* Feb. 26, 2022, www.sfchronicle.com/projects/2022/visuals/san-franciso-tore-down-embarcadero-freeway/.

52 *San Andreas Fault ruptured:* "The 1989 Loma Prieta Earthquake," California Department of Conservation, www.conservation.ca.gov/cgs/earthquakes/loma-prieta.

52 *"the opportunity of a lifetime":* Andrew Chamings, "'A Monstrous Mistake': Remembering the Ugliest Thing San Francisco Ever Built," SFGATE, March 3, 2021, www.sfgate.com/local/editorspicks/article/embarcadero-freeway-san-francisco-photos-history-15990662.php.

52 *Transit ridership increased:* "Embarcadero Freeway," Congress for the New Urbanism, www.cnu.org/what-we-do/build-great-places/embarcadero-freeway.

53 *board of supervisors voted:* Arredondo, "Rise and Demise of San Francisco's 'Hideous' Embarcadero Freeway."

53 *the West Side Highway:* Peter Simek, "What Other Cities Learned," *D Magazine,* April 22, 2014, www.dmagazine.com/publications/d-magazine/2014/may/what-other-cities-learned-tearing-down-highways/2/.

53 *"gleaming new concrete ribbon":* Victor H. Bernstein, "West Side Highway to Open," *New York Times,* Oct. 10, 1937.

53 *Portland's Harbor Drive:* "Model Cities: Portland—Harbor Drive," Congress for the New Urbanism, www.cnu.org/highways-boulevards/model-cities/portland.

53 *Milwaukee's mayor, John Norquist:* "Model Cities: Milwaukee—Park East Freeway," Congress for the New Urbanism, www.cnu.org/highways-boulevards/model-cities/milwaukee.

54 *active campaigns:* "Highways to Boulevards Fact Sheet," Congress for the New Urbanism, www.cnu.org/our-projects/highways-boulevards.

54 *eighteen North American cities:* Ibid.

54 *In 1998, a global study:* Phil Goodwin et al., "Evidence on the Effects of Road Capacity Reduction on Traffic Levels," *Journal of Transportation Engineering and Control,* 1998, 348–54.

54 *"even after controlling":* Robert Cervero et al., "From Elevated Freeways to Surface Boulevards: Neighborhood and Housing Price Impacts in San Francisco," *Journal of Urbanism,* 2009, www.tandfonline.com/doi/full/10.1080/17549170902833899.

54 *more than $1 billion:* "Project Profile: Park East Freeway Removal Project, Milwaukee, Wisconsin," U.S. Department of Transportation, www.fhwa.dot.gov/ipd/project_profiles/wi_park_east_freeway.aspx.

54 *In Dallas, a 2021 study:* "I-345/45 Framework Plan," Toole Design Group, April 14, 2021, www.coalitionforanewdallas.org/featured-posts/toole-framework-plan.

54 *Claiborne Expressway:* Audra D. S. Burch, "One Historic Black Neighborhood's

Stake in the Infrastructure Bill," *New York Times,* Nov. 20, 2021, www.nytimes
.com/2021/11/20/us/claiborne-expressway-new-orleans-infrastructure.html.

54 *hundreds of sprawling live oaks:* Katy Reckdahl, "A Divided Neighborhood Comes
Together Under an Elevated Expressway," Next City, Aug. 20, 2018, nextcity.org/
features/a-divided-neighborhood-comes-together-under-an-elevated-expressway.

55 *100,000 cars a day:* "Embarcadero Freeway," Congress for the New Urbanism.

56 *Elm Street was the heart:* Paula Bosse, "The Gypsy Tea Room, Central Avenue, and
the Darensbourg Brothers," *Flashback Dallas* (blog), Dec. 20, 2015, flashbackdallas
.com/2015/12/20/gypsy-tea-room/.

56 *cultural and commercial heart:* Jay Brakefield and Alan B. Govenar, *Deep Ellum and
Central Track: Where the Black and White Worlds of Dallas Converged* (Denton: University of North Texas Press, 1998), xv–xix.

56 *By the 1940s:* Michael Phillips, *White Metropolis: Race, Ethnicity, and Religion in Dallas,
1841–2001* (Austin: University of Texas Press, 2006), 65.

56 *Knights of Pythias Temple:* "Knights of Pythias Temple (Also Known as the Union
Bankers Building)," Dallas Landmark Structures and Sites, City of Dallas (2018),
dallascityhall.com/departments/sustainabledevelopment/historicpreservation/
Pages/knights_of_pythias_temple.aspx.

57 *Construction of I-345:* "State Agency OK's Work on Elevated Road."

57 *consumed the 2400 block:* Dallas City Directory, 1950–1975, microfilm, Dallas Public
Library.

57 *fifty-four city blocks:* "Remove," Coalition for a New Dallas, www.coalitionforanew
dallas.org/remove.

57 *purchased 133 properties:* Analysis by author of Texas Department of Transportation
Right of Way Division Records, Texas Digital Archive.

58 *"hostile to remembrance":* Robin Young, "In the 1920s, 1 in 3 Eligible Men in Dallas
Were KKK Members," *Here & Now,* WBUR, Nov. 6, 2019, www.wbur.org/
hereandnow/2019/11/06/ku-klux-klan-dallas-texas.

58 *the accommodation that crafted:* Schutze, *Accommodation.*

58 *"In this obsessively image-conscious city":* Phillips, *White Metropolis,* 3.

Chapter 6: Expand

59 *metropolitan planning organizations:* "Urban Transportation Planning in the United
States: An Historic Overview," U.S. Department of Transportation (1988), ampo
.org/about-us/about-mpos/.

60 *convened to consider:* Houston-Galveston Area Council Transportation Policy Council public meeting, Items 3 and 8, July 16, 2019, hgac.swagit.com/play/07262019-568.

60 *largest road project:* Dug Begley, "Houston Leaders Asked to Commit $100M to Mas-
sive I-45 Rebuild," *Houston Chronicle,* July 7, 2019, www.houstonchronicle.com/
news/houston-texas/transportation/article/Houston-leaders-asked-to-commit
-100M-to-massive-14073908.php.

60 *draft environmental impact statement:* "Draft Environmental Impact Statement,
North Houston Highway Improvement Project," Texas Department of Transportation, April 2017, 3-11 to 3-15.

60 *expand by 50 percent:* "NHHIP Community Workshops: Virtual Meeting," City of
Houston Planning and Development Department, Oct. 2020, 22.

60 *220 feet across to 570:* Matthew Casale et al., "Highway Boondoggles 5," U.S. Public
Research Interest Group Education Fund, June 2019, 18.

61 *"all alternatives would cause":* "Draft Environmental Impact Statement," ES-14.

63 *elected as the chief executive:* Blake Patterson, "She's 28. She's an Immigrant. She's in
Charge of Texas' Most Populous County. Get Used to It," *Texas Observer,* April 3,
2019, www.texasobserver.org/lina-hidalgo-harris-county-judge/.

64 *The council approves:* Chloe Alexander, "Update: $100M Preliminary Funding Approved for I-45 Expansion Project Despite Protest," *Houston Business Journal,* July 26, 2019, www.bizjournals.com/houston/news/2019/07/26/mattress-mack -judge-lina-hidalgo-join-dozens-in.html.

66 *a brief stint as Enteron:* Holman W. Jenkins, Jr., "Enron Is History, Says History," *Wall Street Journal,* Nov. 28, 2001.

68 *George P. Mitchell bought:* Roger Galatas and James Barlow, *The Woodlands: The Inside Story of Creating a Better Hometown* (Chicago: Urban Land Institute, 2004), xi.

68 *"We're stuffing up the highways":* "Oilman Mitchell Behind 'New Town,'" *Houston Post,* Dec. 21, 1972.

68 *"a new hometown":* The Woodlands sales brochure, circa 1974, Houston Public Library.

69 *grand opening:* "The Woodlands Opens Saturday!," *Houston Chronicle,* Oct. 20, 1974.

69 *more persistent reality:* William Pack, "Jobs Top Priority of Woodlands," *Houston Post,* June 23, 1985.

69 *"blacks were not inherently opposed":* Ibid.

69 *yet another suburb:* Loren Steffy, *George P. Mitchell: Fracking, Sustainability, and an Unorthodox Quest to Save the Planet* (College Station: Texas A&M University Press, 2019), 287.

71 *a new method for extracting:* Ibid., 224–28.

Chapter 7: Again

72 *The tape spans 107 feet:* Heyden Black Walker, "Austin Has the Chance to Fix This Deadly Barrier," Active Towns, Sept. 1, 2021, www.youtube.com/watch?v= c9nbkAnoa8k.

73 *TxDOT published schematics:* I-35 Capital Express Central Virtual Public Meeting, Aug. 10, 2021, Texas Department of Transportation, my35capex.com/events/683/.

73 *twenty-lane highway:* I-35 Capital Express Central Virtual Public Meeting: Proposed Build Alternatives Schematics and Layers, Texas Department of Transportation.

73 *TxDOT intended to take:* "Draft Alternatives Evaluation Technical Report," Texas Department of Transportation, Aug. 2021, 20.

73 *"We're not pretending to say":* Austin Sanders, "Council Gets Briefed on Controversial I-35 Expansion Plans," *Austin Chronicle,* Sept. 3, 2021, www.austinchronicle.com/ news/2021-09-03/council-gets-briefed-on-controversial-i-35-expansion-plans/.

74 *flurry of bills:* "Series of New Bills at Texas Legislature Focus on Transportation Projects," Spectrum News, Feb. 27, 2019, spectrumlocalnews.com/tx/austin/ news/2019/02/27/series-of-new-bills-at-texas-legislature-focus-on-transportation.

74 *"There is no highway fairy":* Andrew Wilson, "Texas Sen. Kirk Watson Files Bills to Ease I-35 Traffic in Travis County," KVUE, Feb. 25, 2019, www.kvue.com/article/ news/texas-sen-kirk-watson-files-bills-to-ease-i-35-traffic-in-travis-county/269 -c386551f-03ce-4762-af4b-6d0e4af19d09.

76 *downtown streets master plan:* "Downtown Great Streets Master Plan," prepared by Black & Vernooy and Kinney & Associates for the City of Austin, Nov. 2001.

76 *three city blocks:* Aaron Seward, "In the Black," *Texas Architect,* May/June 2017, magazine.texasarchitects.org/2017/05/16/in-the-black/.

77 *towering off-ramps:* Chuck Lindell, "I-35 Underground?," *Austin American-Statesman,* June 9, 1996.

77 *But in 2011, TxDOT returned:* "Draft Environmental Impact Statement: I-35 Capital Express Central Project from US 290 East to US 290 West/SH 71," Texas Department of Transportation, Dec. 2022, 12–13.

77 *reclaim thirty acres:* "Reimagining I-35," Reconnect Austin, Nov. 2020, 16, issuu .com/reconnectaustin/docs/2020-06_reconnect-benefits_single-p _b7668973670e66.

78 *at least sixty thousand units:* "Austin Strategic Housing Blueprint," City of Austin (2017), www.austintexas.gov/page/view-blueprint.

79 *State Highway 130:* "Consider the Recommended Designation Switch," Reconnect Austin, Oct. 28, 2019.

79 *roughly 97 percent:* "Transportation Funding in Texas," Texas Department of Transportation (2019), 5.

79 *investment in public transportation:* Mike Clarke-Madison, "From Light Rail to a Downtown Tunnel: The Parts of Project Connect," *Austin Chronicle,* Sept. 25, 2020, www.austinchronicle.com/news/2020-09-25/from-light-rail-to-a-downtown -tunnel-the-parts-of-project-connect/.

79 *Federal funding:* Samuel King, "What's Project Connect's First Stop After Austin Voters Handily Pass Prop A?," KUT, Nov. 4, 2020, www.kut.org/transportation/2020 -11-04/whats-project-connects-first-stop-after-austin-voters-handily-pass-prop-a.

80 *three "community concepts":* "Evaluation of TxDOT Build Alternatives and Community Concepts," Texas A&M Transportation Institute, Aug. 2021, 26.

81 *Dina Flores opened:* "Our Story," Escuelita del Alma, escuelitadelalma.com/about/.

81 *very different place:* Mary Huber, "Fast Forward: Austin Metro Area Sees Two Decades of Explosive Growth," *Austin American-Statesman,* Jan. 10, 2020, stories .usatodaynetwork.com/2020-vision-austin/austin-metro-area-sees-two-decades-of -explosive-growth/.

82 *"Ladies and gentlemen":* Gary Cartwright, "Tex-Mex and the City," *Texas Monthly,* Feb. 2007, www.texasmonthly.com/articles/tex-mex-and-the-city/.

82 *city council wrote:* "Our Story," Escuelita del Alma.

Chapter 8: Pause

87 *gentrification had swept:* Monique Welch, "These Houston Neighborhoods Are Changing Through Gentrification. Here's a Look at Their Past and Present," *Houston Chronicle,* Sept. 17, 2021, www.houstonchronicle.com/projects/2021/visuals/ houston-evolving-neighborhoods-gentrification-census/.

87 *giant swath of land:* Mike Snyder, "Moving Up or Pushing Out? How One Massive Project Could Reshape a Community," *Houstonia Magazine,* May 26, 2021, www .houstoniamag.com/news-and-city-life/2021/05/east-river-project-houston-fifth -ward-gentrification.

88 *Sean Jefferson lived:* "Community Voices Along the I-45 Corridor," LINK Houston, i45expansionimpacts.org/sean-jefferson.html.

88 *the only neighborhood:* "NHHIP Community Workshops: Virtual Meeting," City of Houston Planning and Development Department, Oct. 2020, 41–42.

89 *Turner presented:* Turner to Laura Ryan (Texas transportation commissioner), May 12, 2020.

89 *Turner again sent:* Turner to Eliza Paul (TxDOT Houston district engineer), Dec. 8, 2020.

89 *memorandum of understanding:* Emma Whalen, "H-GAC Group Stalls I-45 Resolution After TxDOT Objects to Agreement," Community Impact, Jan. 25, 2021, communityimpact.com/houston/heights-river-oaks-montrose/transportation/ 2021/01/25/h-gac-group-stalls-i-45-resolution-after-txdot-objects-to-agreement/.

89 *National Environmental Policy Act:* ceq.doe.gov.

90 *own NEPA compliance:* "Memorandum of Understanding Between the Federal Highway Administration and the Texas Department of Transportation Concerning State of Texas' Participation in the Project Delivery Pursuant to 23 U.S.C. 327," Dec. 2014, www.txdot.gov/about/programs/environmental/nepa-assignment -documentation.html.

90 *NEPA Assignment:* "NEPA Assignment," Center for Environmental Excellence, American Association of State Highway and Transportation Officials, environment .transportation.org/laws-agreements/nepa/nepa-assignment/.

90 *TxDOT issued:* "TxDOT Announces Record of Decision," Texas Department of Transportation, Feb. 4, 2021, www.txdot.gov/content/dam/project-sites/nhhip/ docs/txdot-announces-rod.pdf.

90 *Menefee had campaigned:* Delger Erdensanaa, "Christian Menefee Aims to Bring Polluters to Justice," *Texas Observer,* Oct. 5, 2022, www.texasobserver.org/christian -menefee-pollution-union-pacific/.

90 *"ignored serious harms":* Harris County, Texas v. Texas Department of Transportation, Case 4:21-cv-00805, filed March 11, 2021.

91 *convened a press conference:* "Watch Live: Harris County Judge Hidalgo Addresses I-45 Expansion Lawsuit," KHOU, March 11, 2021, www.youtube.com/watch?v= tuGvgLV8FWk.

92 *her own complaint:* Nelson to Bass, Jan. 18, 2021.

92 *On March 8:* Achille Alonzi (FHWA division administrator) to James Bass (TxDOT executive director), March 9, 2021.

92 *"a really huge deal":* Sam Mintz, "DOT Halts Texas Highway Project in Test of Biden's Promises on Race," *Politico,* April 1, 2021, www.politico.com/news/2021/ 04/01/dot-texas-highway-equity-478864.

93 *Del Richardson & Associates:* "Happy Clients," Del Richardson & Associates, Inc., www.drainc.com/happy-clients.

93 *had meant stop:* Achille Alonzi (FHWA division administrator) to Marc Williams (TxDOT executive director), June 14, 2021.

Chapter 9: Transit

97 *$2.3 trillion infrastructure plan:* "Remarks by President Biden on the American Jobs Plan," Carpenters Pittsburgh Training Center, March 31, 2021, www.whitehouse .gov/briefing-room/speeches-remarks/2021/03/31/remarks-by-president-biden -on-the-american-jobs-plan/.

97 *a historic investment:* "Fact Sheet: The American Jobs Plan," White House, March 31, 2021, www.whitehouse.gov/briefing-room/statements-releases/2021/ 03/31/fact-sheet-the-american-jobs-plan/.

97 *convenes a virtual hearing:* "21st Century Communities: Public Transportation Infra- structure Investment and FAST Act Reauthorization," Senate Committee on Bank- ing, Housing, and Urban Affairs, April 15, 2021, www.banking.senate.gov/ hearings/21st-century-communities-public-transportation-infrastructure -investment-and-fast-act-reauthorization.

98 *This is not true:* Associated Press, "Money for Highway Trust Fund," *New York Times,* Sept. 11, 2008, www.nytimes.com/2008/09/11/us/11brfs -MONEYFORHIGH_BRF.html.

98 *more than $140 billion:* "What Is the Highway Trust Fund, and How Is It Financed?," Tax Policy Center, Urban Institute and Brookings Institution, May 2020, www .taxpolicycenter.org/briefing-book/what-highway-trust-fund-and-how-it-financed.

100 *more than thirty Democrats:* "Reps. García, Pressley, and Jeffries Introduce Transit Parity Resolution," U.S. Representative Jesús G. "Chuy" García, Dec. 10, 2020, chuygarcia.house.gov/media/press-releases/reps-garcia-pressley-and-jeffries -introduce-transit-parity-resolution.

100 *for-profit enterprise:* George Smerk, *The Federal Role in Urban Mass Transportation* (Bloomington: Indiana University Press, 1991), 34–37, 47, 54.

100 *The first transit system:* Jackson, *Crabgrass Frontier,* 33–34.

100   *"Mass transit was a centerpiece":* Smerk, *Federal Role in Urban Mass Transportation,* 37.

100   *Transit ridership peaked:* "1987 Transit Fact Book," American Public Transit Association, 31–32.

101   *"Because it had always been":* Smerk, *Federal Role in Urban Mass Transportation,* 54.

101   *"These passengers aren't lost":* Ann Dear, "We Should Open Up the Highway Trust Fund Now," *Reader's Digest,* entered into Congressional Record, Senate, April 4, 1973, 10987 (p. 26).

102   *"highway-auto-petroleum complex":* William V. Shannon, "The Untrustworthy Highway Fund," *New York Times Magazine,* Oct. 15, 1972.

102   *"Critics of the automobile's long supremacy":* Ibid.

102   *In 1972, for every dollar:* Dear, "We Should Open Up the Highway Trust Fund Now."

102   *Cooper-Muskie amendment:* William V. Shannon, "The Highwaymen," *New York Times,* Sept. 17, 1972.

102   *"This bill would correct":* Congressional Record, U.S. Senate Committee on Public Works, Jan. 18, 1973, 88–92.

103   *raise property taxes:* "A History of BART: The Concept Is Born," Bay Area Rapid Transit, www.bart.gov/about/history.

103   *John B. Anderson:* Congressional Record, House, Oct. 2, 1972, 33427 (p. 71).

104   *"there was an element of desperation":* Smerk, *Federal Role in Urban Mass Transportation,* 117.

104   *a diversion of $800 million:* Ibid., 119.

104   *"The Act is not only a highway act":* Ibid., 118.

104   *highway spending surpassed:* "Public Law 93-87—August 13, 1973," Federal-Aid Highway Act.

105   *gas tax hadn't been raised:* "Memorandum: Historical Table of Federal Excise Tax Rates on Gasoline," Congressional Research Service.

105   *gas tax to ten cents:* Jeff Davis, "Reagan Devolution: The Real Story of the 1982 Gas Tax Increase," Eno Center for Transportation, Sept. 2015.

105   *"Right now there is":* Ibid., 12.

106   *"We cannot leave":* Ibid., 20.

106   *remained essentially flat:* Email from Jeff Davis (Eno Center for Transportation), Sept. 27, 2022.

107   *status quo remains intact:* "Understanding the 2021 Infrastructure Law," Transportation for America, t4america.org/iija/#start.

108   *formula funding system:* Jeff Davis, Alice Grossman, and Paul Lewis, "Refreshing the Status Quo: Federal Highway Programs and Funding Distribution," Eno Center for Transportation, Oct. 2019, 6–7.

108   *$1.5 billion program:* "TIGER Discretionary Grant Program," U.S. Department of Transportation, www.transportation.gov/highlights/tiger/tiger-discretionary -grant-program.

109   *portion of its Inner Loop:* Inner Loop East Project, City of Rochester, www .cityofrochester.gov/InnerLoopEast/.

109   *"like a noose":* Zack Seward, "Rochester Doubling Down on Inner Loop Plans," WXXI, Feb. 14, 2012, www.innovationtrail.org/money/2012-02-14/rochester -doubling-down-on-inner-loop-plans.

109   *tearing down the Claiborne Expressway:* "Final Report: Charting the Future of the Claiborne Communities," prepared by Kittelson & Associations Inc. and Goody Clancy for City of New Orleans, March 2014.

109   *Sheridan Expressway:* Noah Kazis, "TIGER II Funds Sheridan Replacement Study, Fordham Redesign," *Streetsblog,* Oct. 15, 2010, nyc.streetsblog.org/2010/10/15/ tiger-ii-funds-sheridan-replacement-study-fordham-redesign/.

109   *"reconnect neighborhoods":* "Fact Sheet: The American Jobs Plan," White House

Briefing Room, March 31, 2021, www.whitehouse.gov/briefing-room/statements
-releases/2021/03/31/fact-sheet-the-american-jobs-plan/.

110 *Amy talked to:* Burch, "One Historic Black Neighborhood's Stake in the Infrastructure Bill."

110 *$3 billion in the House bill:* H.R. 3684, 117th Cong., introduced June 4, 2021, Sec. 1101, www.congress.gov/bill/117th-congress/house-bill/3684/text/ih #H48C57CA6426049988D4C607F522470B6.

110 *Duluth Waterfront Collective:* Dan Kraker, "Freeway Fighter: A Vision to Replace I-35 in Duluth Gains Momentum," MPR, Jan. 11, 2022, www.mprnews.org/story/2022/01/11/freeway-fighter-a-vision-to-replace-i35-in-duluth-gains-momentum.

111 *Freeway Fighters Network:* "Reconnecting Communities Coalition Letter," Aug. 19, 2021, freeway-fighters.org/2021/08/19/reconnecting-communities-coalition-letter/.

112 *"This law makes this":* "Remarks by President Biden at Signing of H.R. 3684, the Infrastructure Investment and Jobs Act," South Lawn, White House, Nov. 15, 2021, www.whitehouse.gov/briefing-room/speeches-remarks/2021/11/15/remarks-by -president-biden-at-signing-of-h-r-3684-the-infrastructure-investment-and-jobs-act/.

112 *Infrastructure Investment and Jobs Act:* "Show Me the Money: Financial Breakdown of the Infrastructure Bill," Transportation for America, Dec. 15, 2021, t4america.org/2021/12/15/show-me-the-money-financial-breakdown-of-the-infrastructure-law/.

Chapter 10: Hemispheres

113 *five options for the future:* Virtual Public Meeting for I-345 Feasibility Study in Dallas County, Texas Department of Transportation, June 22, 2021, www.keepitmoving dallas.com/I345_archive.

114 *replace the car capacity:* Patrick Kennedy, "Where the 345 Feasibility Process Went Off the Rails," Medium, June 27, 2022, medium.com/@WalkableDFW/where-the -345-feasibility-process-went-off-the-rails-fb088568cc1d.

114 *current-day travel patterns:* "I-345 Feasibility Study: Origin and Destination Traffic Data," Texas Department of Transportation, June 2021.

115 *"High-level modeling":* "I-345 Feasibility Study: Traffic Volume Analysis," Texas Department of Transportation, June 2021.

115 *two unequal hemispheres:* "Comprehensive Housing Policy Racial Equity Assessment," prepared by TDA Consulting for the City of Dallas Department of Housing and Neighborhood Revitalization, Oct. 25, 2021, 4, 19–26.

115 *legalize housing segregation:* J. H. Cullum Clark, "Equitable Development in Southern Dallas," Dallas Collaboration for Equitable Development, Jan. 2021, 9.

115 *Ku Klux Klan:* Darwin Payne, "When Dallas Was the Most Racist City in America," *D Magazine*, May 22, 2017, www.dmagazine.com/publications/d-magazine/2017/june/when-dallas-was-the-most-racist-city-in-america/.

115 *In 1937, when an appraiser:* Robert K. Nelson et al., "Mapping Inequality," in Nelson and Ayers, *American Panorama*, dsl.richmond.edu/panorama/redlining/#loc=12/32.784/-96.884&city=dallas-tx.

116 *Only one thousand were available:* Lewis, "Under Asphalt and Concrete," 107.

116 *"There is no room":* Ibid., 92.

116 *a series of bombings:* Schutze, *Accommodation*, 25–28.

116 *twenty-two years longer:* Ibid., 6.

116 *64 percent of the landmass:* Clark, "Equitable Development in Southern Dallas," 8.

116 *17 percent of its jobs:* Joel Kotkin and J. H. Cullum Clark, "Big D Is a Big Deal," *City Journal* (Summer 2021).

116 *average commute times:* "Fraction with Short Work Commutes in 2012–16," The Opportunity Atlas: Dallas, Texas, www.opportunityatlas.org.

116 *household location by race:* "North Texas Regional Housing Assessment," City of Dallas, Nov. 2018, 64.

122 *FOR SALE sign:* Peter Simek, "Reclaiming the Forest Theater," *D Magazine,* Sept. 29, 2017, www.dmagazine.com/publications/d-magazine/2017/october/forest -theater-south-dallas/.

122 *students planted peas:* "One College Turns Its Football Field into a Farm and Sees Its Students Transform," *PBS NewsHour,* Sept. 12, 2016, www.pbs.org/newshour/ show/one-college-turns-football-field-farm-sees-students-transform.

123 *lease the theater:* Robert Wilonsky, "Erykah Badu's out of the Black Forest, but the Theater's Owners Swear a Rebirth by Summer," *Dallas Observer,* Jan. 14, 2009.

123 *a large chalkboard:* Simek, "Reclaiming the Forest Theater."

123 *announced a plan:* Sriya Reddy, "South Dallas Nonprofit Unveils Plans for Reopening of Forest Theater," *Dallas Morning News,* Nov. 5, 2021, www.dallasnews.com/news/ 2021/11/05/south-dallas-nonprofit-unveils-plans-for-the-reopening-of-forest-theater/.

124 *Only 25 percent:* Clark, "Equitable Development in Southern Dallas," 10–11.

124 *Some blocks in South Dallas:* Jim Schutze, "Want to Make Real Money on Real Estate? Go to South Dallas," *Dallas Observer,* Aug. 29, 2019, www.dallasobserver.com/ news/hottest-property-investments-in-dallas-are-in-south-dallas-11743541.

125 *the school reopened:* Skye Seipp, "A $7 Million Facelift Will Transform South Dallas' MLK School into an Arts Academy," *Dallas Free Press,* July 8, 2020, dallasfreepress .com/south-dallas/a-7-million-facelift-will-transform-south-dallas-mlk-school-into -an-arts-academy/.

125 *renamed S. M. Wright Freeway:* Oscar Slotboom, *Dallas–Fort Worth Freeways: Texas- Sized Ambition* (self-pub., 2014), 295–97.

126 *Dead Man's Curve:* Bruce Tomaso, "Dallas Officials Urge Rebuilding Wright Free- way," *Dallas Morning News,* Feb. 8, 2008.

126 *held a public meeting:* Daniel Rodrigue, "TxDOT Dazzles Southern Dallas with Plans to Straighten Out 'Statistically Dangerous Curve' on S. M. Wright," *Dallas Observer,* March 31, 2010, www.dallasobserver.com/news/txdot-dazzles-southern-dallas -with-plans-to-straighten-out-statistically-dangerous-curve-on-sm-wright-7148487.

126 *TxDOT's version of a boulevard:* "Some Say Fix for S. M. Wright Freeway Isn't Enough," *Dallas Morning News,* May 16, 2010, www.dallasnews.com/news/2010/ 05/16/some-say-fix-for-s-m-wright-freeway-isn-t-enough/.

127 *Trinity Parkway toll road:* Brandon Formby, "How Angela Hunt Changed Dallas' Mind About Its Divisive Riverside Toll Road," *Texas Tribune,* Aug. 9, 2017, www .texastribune.org/2017/08/09/how-angela-hunt-changed-dallas-mind-about-its -riverside-toll-road/.

127 increase *traffic on nearby highways:* "Federal Study: Trinity Toll Road Will Increase Traffic on Other Major Roads," *Dallas Morning News,* April 8, 2015, www .dallasnews.com/news/transportation/2015/04/08/federal-study-trinity-toll-road -will-increase-traffic-on-other-major-roads/.

Chapter 11: Uncertainty

129 *threatened to pull funding:* Texas Transportation Commission public meeting, June 30, 2021, txdot.new.swagit.com/videos/06302021-678.

129 *released on SurveyMonkey:* "TxDOT's Way or No Highway? SurveyMonkey Poll No Way to Decide Future of I-45," editorial, *Houston Chronicle,* Aug. 3, 2021, www .houstonchronicle.com/opinion/editorials/article/Editorial-TxDOT-s-way-or-no -highway-16358946.php.

133 *Mattress Mack refunded:* Fernando Alfonso III, "Mattress Mack Loses $10 Million on World Series Bet," Chron.com, Nov. 2, 2017, www.chron.com/sports/astros/ article/Mattress-Mack-world-series-bet-10-million-astros-12325882.php.

134 *supporter of the Tea Party:* Matthew Tresaugue, "Woodlands Tea Party Group Seeks to Give Cruz a Boost—Again," *Houston Chronicle,* Feb. 27, 2016, www .houstonchronicle.com/politics/article/Woodlands-tea-party-group-seeks-to-give -Cruz-a-6858545.php.

134 *whether to allow state highway funds:* H.J.R. 109, Texas State Legislature, Sess. 87(R), capitol.texas.gov/BillLookup/History.aspx?LegSess=87R&Bill=HJR109.

136 *The results of the survey:* Texas Transportation Commission public meeting, Aug. 31, 2021, txdot.new.swagit.com/videos/08312021-654.

136 *just over 8,000:* Dug Begley, "Controversial I-45 Project Still Alive as TxDOT, Feds Given 90 Days to Settle Differences," *Houston Chronicle,* Aug. 31, 2021, www .houstonchronicle.com/news/houston-texas/transportation/article/I-45-project -still-on-the-table-with-90-day-16425828.php.

136 *viable path forward:* Ibid.

136 *"Chairman Bugg misrepresented my position":* Houston City Council public meeting, Sept. 14, 2021, houstontx.new.swagit.com/videos/159366 (clip: www.instagram .com/p/CT3FvRXA1U_/).

137 *fifty people gather:* "Congresswoman Sheila Jackson Lee Invites Stephanie Pollack, Deputy Administrator of the Federal Highway Administration, to Houston to Address Major Concerns with the I-45 Project," Congresswoman Sheila Jackson Lee, Facebook live, www.facebook.com/watch/live/?ref=watch_permalink&v= 620425962434543; www.facebook.com/watch/live/?ref=watch_permalink&v= 481454376638446.

138 *purchased the property:* Harris County Appraisal District Real Property Account Information, Property No. 0371080090065.

138 *Bruce Elementary School:* "Health Impact Assessment of the North Houston Highway Improvement Project," Air Alliance Houston, May 2019, 48, app. 3.

138 *Independence Heights:* "Super Neighborhood 13—Independence Heights," City of Houston, www.houstontx.gov/superneighborhoods/55.html.

138 *spliced the community:* Tanya Debose, "Houston's Historic Independence Heights' Complicated Past, Present, and Future with Divisive Freeways," Kinder Institute for Urban Research, June 14, 2019, kinder.rice.edu/urbanedge/houstons-historic -independence-heights-complicated-past-present-and-future-divisive.

Chapter 12: Demand

141 *inserted an amendment:* Dug Begley, "Houston Congressmen Again Battle over Richmond Rail Funding," *Houston Chronicle,* June 9, 2014.

142 *"It's a throwback":* House of Representatives, floor debate on the Transportation-Housing and Urban Development Appropriations bill, June 9, 2014, www.c-span .org/video/?319859-2/house-session-part-1&start=21678&transcriptSpeaker= 1011398.

142 *summarized his findings:* Jay Blazek Crossley, "It Took 51% More Time to Drive Out Katy Freeway in 2014 Than in 2011," Houston Tomorrow, May 26, 205, accessed via Wayback Machine, web.archive.org/web/20150530015933/http://www .houstontomorrow.org/livability/story/it-took-51-more-time-to-drive-out-katy -freeway-in-2014-than-2011/.

142 *Culberson's amendment remained:* H.R. 4745, 113th Cong., 2nd Sess., 52.

143 *construction of the Grand Parkway:* "SH 99/Grand Parkway Project," Texas Department of Transportation, www.txdot.gov/projects/projects-studies/houston/sh99 -grand-parkway.html.

143 *real estate development:* Jen Rice, "Houston Keeps Paving over Rain-Absorbent Katy Prairie, Even After Devastating Harvey Impacts," *Houston Chronicle,* Aug. 24, 2022.

143 *zoned for single-family housing:* "Land Development Draft Code & Map," City of

Austin, Jan. 31, 2020, www.austintexas.gov/department/land-development-draft
-code-map#maps.

143 *exploded in population:* "Demographics," City of Kyle Economic Development,
kyleed.com/local-data/demographics.

145 *exclusionary zoning code:* Megan Kimble, "The Fight to Make Austin Affordable,"
Texas Observer, Dec. 5, 2019, www.texasobserver.org/the-fight-to-make-austin
-affordable-housing/.

145 *"misinformation, hyperbole, fearmongering":* Steve Adler, "We Should Consider if We
Need a New and Different Process to Fix Our Land Development Code," City of
Austin Council Message Board, Aug. 1, 2018.

146 *The city estimated:* "Land Development Code Revision: Report Card," City of Austin, Feb. 21, 2020.

146 *homeowners sued the city:* Megan Kimble, "Desperate for Housing, Austin Seeks Relief in Rezoning," Bloomberg, April 29, 2022, www.bloomberg.com/news/
features/2022-04-29/as-gentrification-sweeps-austin-zoning-reform-remain-elusive.

147 *Dewitt C. Greer State Highway Building:* "1918 State Office Building and 1933 State
Highway Building," National Register of Historic Places, National Park Service.

147 *art deco structure:* Bud Franck, "Dewitt Greer State Highway Building," Guide to
Austin Architecture, guidetoaustinarchitecture.com/places/dewitt-c-greer-state
-highway-building/.

148 *rural paved roads:* Aman Batheja, "TxDOT Ends Program That Converts Paved
Roads to Gravel," *Texas Tribune,* Oct. 24, 2014.

148 *"no longer needed":* Minute Order, Texas Department of Transportation, Dec. 18,
2014.

148 *Broadway was a "stroad":* Charles Marohn, "The STROAD," Strong Towns,
March 4, 2013, www.strongtowns.org/journal/2013/3/4/the-stroad.html.

149 *Broadway redevelopment plan:* Shari Biediger, "TxDOT Votes to Take Control of
Broadway, Short-Circuiting City's Redevelopment Plans," *San Antonio Report,*
Jan. 27, 2022, sanantonioreport.org/txdot-vote-broadway-redevelopment/.

149 *more than $760 million:* "Broadway Corridor," Centro San Antonio,
centrosanantonio.org/the-broadway-corridor/.

149 *"I would like to present":* Texas Transportation Commission public meeting, Jan. 27,
2022, txdot.new.swagit.com/videos/01282022-520.

150 *phone call from Abbott:* Greg Jefferson, "In State's Broadway U-Turn, San Antonio
Banker Bugg Said It but Abbott Was at the Wheel," *San Antonio Express-News,*
Jan. 28, 2022, www.expressnews.com/business/business_columnists/greg
_jefferson/article/Jefferson-In-state-s-Broadway-U-turn-San-16813560.php.

150 *synonymous with economic development:* "Texas Transportation Commission Chairman J. Bruce Bugg, Jr. Interviews Texas Governor Greg Abbott," Texas Transportation e-forum, Feb. 9, 2021, www.youtube.com/watch?v=XNWt9lLT8Fk.

150 *governor's top campaign donors:* Patrick Svitek et al., "In Texas, Where Money Has
Long Dominated Politics, Greg Abbott Is in a League of His Own," *Texas Tribune,*
Oct. 18, 2022, www.texastribune.org/2022/10/18/greg-abbott-texas-fundraising
-governor-donors/.

151 *economic cost of crashes:* Jay Blazek Crossley, "2021 Cost of Crashes in Texas,"
Farm&City.

Chapter 13: Housing

152 *flooded during Hurricane Harvey:* Mike Morris, "Houston's Sprawling Drainage Project Would Help Hundreds of Homes Along White Oak Bayou," *Houston Chronicle,*
March 16, 2018, www.houstonchronicle.com/news/article/Houston-s-sprawling
-drainage-project-would-help-12759536.php.

NOTES 311

152 *Built in 1952:* "Clayton Homes Apartments," Houston Housing Authority, housingforhouston.com/locations/clayton-homes-apartments/.

152 *Clayton Homes to TxDOT:* Dug Begley, "TxDOT's $7 Billion Plan to Shorten Your I-45 Commute May Displace Hundreds of Families," *Houston Chronicle,* June 4, 2020, www.houstonchronicle.com/news/houston-texas/transportation/article/I-45-project-will-let-drivers-step-on-the-gas-15317077.php.

153 *thirty thousand people were on the waiting list:* "Housing Choice Voucher Waitlist," Houston Housing Authority, housingforhouston.com/hcvp-wait-list-status/.

153 *decline to rent:* Edgar Walters and Neena Satija, "Section 8 Vouchers Are Supposed to Help the Poor Reach Better Neighborhoods. Texas Law Gets in the Way," *Texas Tribune,* Nov. 19, 2018, www.texastribune.org/2018/11/19/texas-affordable-housing-vouchers-assistance-blocked/.

153 *struggle to find: Inclusive Communities Project Inc. v. Greg Abbott,* Case 3:17-cv-00440-G, filed Feb. 16, 2017, texashousers.org/2017/02/17/suit-filed-to-overturn-tx-statute-outlawing-municipal-ordinances-to-prohibit-discrimination-against-recipients-of-federal-housing-vouchers/.

155 *Houston's Chinatown:* Jenalia Moreno, "Chinatown No Longer: Call It EaDO, as in 'East Downtown,'" Chron.com, Oct. 17, 2009, www.chron.com/homes/article/Chinatown-no-longer-Call-it-EaDo-as-in-east-1747951.php.

155 *East Downtown Management District:* "Economic Development—TIRZ. East Downtown," City of Houston, www.houstontx.gov/ecodev/tirz/15.html.

155 *luxury urban living:* Lofts at the Ballpark, accessed via Wayback Machine, web.archive.org/web/20200809014037/https://loftsattheballpark.com/#.

156 *"no further actions":* Achille Alonzi (FHWA division administrator) to Marc Williams (TxDOT executive director), June 14, 2021.

156 *relocation assistance from TxDOT:* Suzy Romoser-Broadway (HDR Engineering Inc.) to Winebar, April 28, 2021.

158 *intended to tear down:* Sam González Kelly, "Plans to Demolish Apartments near Minute Maid Park Sparks Backlash from Sheila Jackson Lee, Activists," *Houston Chronicle,* June 22, 2022, www.houstonchronicle.com/news/houston-texas/houston/article/Plan-to-demolish-apartments-near-Minute-Maid-Park-17255957.php.

158 *TxDOT had purchased:* Harris County Appraisal District Real Property Account Information, Property No. 0011790000001.

159 *listed at 165 units:* "Community Impacts Assessment: North Houston Highway Improvement Project," Texas Department of Transportation, Aug. 2020, table 5-3.

160 *"key events in the acquisition":* Statement from FHWA posted on Twitter by Jay R. Jordan (@JayRJordan), June 21, 2022, 7:09 P.M., twitter.com/jayrjordan/status/1539385076869373954?s=20.

162 *TxDOT demolition permits:* Sam González Kelly, "Houston to Delay Demolition of Apartments near Minute Maid Park as Federal Inquiry Begins," *Houston Chronicle,* June 22, 2022, www.houstonchronicle.com/news/houston-texas/houston/article/Houston-to-delay-demolition-of-apartments-near-17258539.php.

162 *the only illicit activity:* Jay R. Jordan, "TxDOT to Demolish Apartments to Make Way for Controversial I-45 Expansion near Downtown Houston—Despite Federal Hold on Project," Chron.com, June 21, 2022, www.chron.com/news/houston-texas/transportation/article/TxDOT-to-demolish-apartments-to-make-way-for-17255413.php.

162 *"You can't take more":* Jay R. Jordan, "Turner Calls on TxDOT to Pledge 'Net-Zero Housing Loss' During I-45 Expansion, but Agency Won't Commit," Chron.com, June 22, 2022, www.chron.com/news/houston-texas/transportation/article/Turner-disappointed-with-Lofts-demolition-asks-17258334.php.

162 *Turner writes that TxDOT:* "Mayor Turner's Statement on the Texas Department of

Transportation's Expansion of Interstate 45 and Demolition of Apartments," City of Houston, June 23, 2022, www.houstontx.gov/mayor/press/2022/statement -txdot-lofts-ballpark.html.

163 *the commission will vote:* Texas Transportation Commission public meeting, Aug. 30, 2022, txdot.new.swagit.com/videos/179686.

163 *unprecedented $85 billion:* "Governor Abbott, TxDOT, Announce Record $85 Billion 10-Year Transportation Plan," Texas Department of Transportation, Aug. 30, 2022, www.txdot.gov/about/newsroom/statewide/85-billion-10year-transportation-plan .html.

164 *civil rights leader in Dallas:* Tim Rogers, "What Does South Dallas Think About Highways? Let's Ask a 'Militant' Black Leader," *D Magazine,* April 9, 2014, www .dmagazine.com/history/2014/04/what-does-south-dallas-think-about-highways -lets-ask-a-militant-black-leader/.

Chapter 14: Access

167 *the "City of Hate":* Wade Goodwyn, "Marking Kennedy Assassination, Dallas Still on 'Eggshells,'" NPR, Nov. 21, 2013, www.npr.org/2013/11/21/246580954/ marking-kennedy-assassination-dallas-still-on-egg-shells.

167 *started with city hall:* Mark Lamster, "Dallas City Hall: Why the City's Most Hated Building Might Be Its Greatest Masterpiece," *Dallas Morning News,* Dec. 21, 2022, www.dallasnews.com/arts-entertainment/architecture/2022/12/21/dallas-city -hall-why-the-citys-most-hated-building-might-be-its-greatest-masterpiece/.

168 *Ceason begins her presentation:* Dallas City Council public meeting, Oct. 19, 2022, dallastx.new.swagit.com/videos/202283.

168 *"With the depressed":* "City Council I-345 Briefing," Texas Department of Transportation, Oct. 2022, 8.

168 *hybrid option:* Ibid., 12.

168 *a highway expansion:* Virtual Public Meeting for I-345 Feasibility Study, Texas Department of Transportation, May 24, 2022, www.keepitmovingdallas.com/I345.

169 *TxDOT wouldn't pay:* Kelsey Thompson, "TxDOT: Austin Has Until Fall 2024 to Identify $350M in Funds for 'Cap and Stitch' Project," KXAN, Feb. 28, 2023, www .kxan.com/news/local/austin/txdot-austin-has-until-fall-2024-to-identify-350m-in -funds-for-cap-and-stitch-project/.

170 *just one minute:* "Dallas City Center Master Assessment Process," Texas Department of Transportation (2016), 191.

170 *"King of Dallas Sprawl":* Peter Simek, "Meet the Kings of Dallas Sprawl," *D Magazine,* March 5, 2019, www.dmagazine.com/publications/d-magazine/2019/march/ meet-the-kings-of-sprawl/.

175 *twenty-six-mile rail line:* Silver Line Project, Dallas Area Rapid Transit, www.dart .org/about/project-and-initiatives/expansion/silver-line-project.

175 *longest rail system:* "DART Celebrates 25 Years of Rail Service in North Texas," DART, June 11, 2021, dart.org/about/news-and-events/newsreleases/newsrelease -detail/dart-celebrates-25-years-of-rail-service-in-north-texas-1591.

175 *kind of coherent plan:* Mark Dent, "Spinning Their Wheels," *Texas Observer,* April 20, 2020, www.texasobserver.org/dart-dallas-public-transit/.

176 *lowest rates of public transit:* Michael Burrows, Charlynn Burd, and Brian McKenzie, "Commuting by Public Transportation in the United States: 2019," American Community Survey Reports, April 2021.

176 *Trains ran faster than buses:* DART Scorecard: Key Performance Indicators, Dallas Area Rapid Transit, www.dart.org/about/about-dart/key-performance-indicator.

177 *launched a total overhaul:* Sharon Grigsby, "In Booming Region, Will DART Overhaul Give Dallas a Bus System That Connects People with Jobs?," *Dallas Morning News,*

Aug. 31, 2021, www.dallasnews.com/news/commentary/2021/08/31/in-booming
-region-will-dart-overhaul-give-dallas-a-bus-system-that-connects-people-with-jobs/.

178 *federal share of transit projects:* H.R. 3684, 117th Cong., introduced June 4, 2021,
sec. 104(c), www.congress.gov/bill/117th-congress/house-bill/3684/text/ih
#H709DF14452D648D6ACC17D0B4CE426C1.

178 *the estimated price:* Memo from David Couch (Project Connect program office) to
Austin Transit Partnership Board, "Update on Program Including Light Rail Project Cost Drivers and Estimates," April 7, 2022.

179 *"I'm not saying it's a perfect storm":* Austin Transit Partnership Board public meeting,
July 20, 2022, austintx.new.swagit.com/videos/177474.

179 *TxDOT would approve:* Dug Begley, "TxDOT Approves $85 Billion, 10-Year Plan to
Widen and Maintain Highways Despite Widespread Opposition," *Houston Chronicle,* Aug. 21, 2022, www.houstonchronicle.com/news/houston-texas/
transportation/article/85-billion-10-year-highway-plan-approved-as-17408289.php.

179 *One of three children:* Dallas Equity Indicators: Poverty, City of Dallas, dallascityhall
.com/departments/pnv/dallas-equity-indicators/Pages/Topic-Details.aspx?topic=
4&theme=1&title=POVERTY.

179 *racial and economic inclusion:* "Inclusive Recovery in U.S. Cities," Urban Institute,
April 2018, 51.

179 *correlation between carlessness and poverty:* David A. King, Michael J. Smart, and Michael Manville, "The Poverty of the Carless: Toward Universal Auto Access," *Journal of Planning Education and Research* 42, no. 3 (2019).

180 *the poorest households:* "The High Cost of Transportation in the United States,"
Transport Matters, Institute for Transportation and Development Policy, May 23,
2019, www.itdp.org/2019/05/23/high-cost-transportation-united-states/.

Chapter 15: Small

184 *agency had released maps:* Nathan Bernier, "Here's Every Property That Would Lose
Land to TxDOT's I-35 Expansion in Central Austin," KUT, Sept. 13, 2021, www.kut
.org/transportation/2021-09-13/heres-every-property-that-would-lose-land-to
-txdots-i-35-expansion-in-central-austin.

184 *maximum amount offered:* "Relocation Assistance," Texas Department of Transportation (2015), 18.

184 *The state required:* "Minimum Standards for Child-Care Centers," Texas Health and
Human Services Commission, March 2023, 216–28.

185 *The pandemic had hit child-care centers:* "Compared to Pre-pandemic Levels, Texas'
Childcare Industry Remains Understaffed," *Texas Standard,* Oct. 17, 2022, www
.texasstandard.org/stories/texas-childcare-industry-understaffed-compared-pre
-covid-levels/.

185 *a survey of 91 child-care centers:* Arezow Doost, "Long Waitlists in Central Texas:
Families Scramble to Find Child Care," KXAN, July 10, 2022, www.kxan.com/
investigations/long-waitlists-in-central-texas-families-scramble-to-find-child-care/.

190 *filed a lawsuit against TxDOT:* Rethink35, Texas Public Interest Research Group, and Environment Texas v. Texas Department of Transportation, Case 1:22-cv-00620, filed
June 26, 2022.

191 *more than thirty new acres:* "Final Environmental Assessment I-35 Capital Express
North Project," Texas Department of Transportation, Dec. 2021, 14; "Final Envi-
ronmental Assessment I-35 Capital Express South Project," Texas Department of
Transportation, Dec. 2021, 10.

191 *Under NEPA:* "Classes of Actions," Code of Federal Regulations, Title 23, Chapter 1, Subchapter H, Part 771 § 771.115, www.ecfr.gov/current/title-23/chapter-I/
subchapter-H/part-771/section-771.115.

191 *TxDOT had issued:* "Finding of No Significant Impact for a FHWA Project," Texas Department of Transportation, I-35 Capital Express North, Dec. 17, 2021; "Finding of No Significant Impact for a FHWA Project," Texas Department of Transportation, I-35 Capital Express South, Dec. 21, 2021.

191 *highway projects can advance in segments:* "Early Coordination, Public Involvement, and Project Development," Code of Federal Regulations, Title 23, Chapter 1, Subchapter H, Part 771 § 771.111, www.ecfr.gov/current/title-23/chapter-I/subchapter -H/part-771/section-771.111.

196 *the reporter includes a quotation:* Nathan Bernier, "Double-Decker Highway Coming to South Austin," KUT, Nov. 15, 2022, www.kut.org/transportation/2022-11-15/ i-35-expansion-south-austin-txdot.

Chapter 16: Repair

202 *A black-and-white aerial photograph:* Seyma Bayram, "The Failed Akron Innerbelt Drove Decades of Racial Inequity. Can the Damage Be Repaired?," *Akron Beacon Journal,* Feb. 3, 2022, www.beaconjournal.com/in-depth/news/2022/02/03/akron -innerbelt-history-racial-inequity-black-history-urban-renewal-ohio/9033520002.

204 *a $17.7 million grant:* "Project Profile: Rochester Inner Loop East, New York, a Freeway to Boulevard," Federal Highway Administration, Center for Innovative Finance Support, www.fhwa.dot.gov/ipd/project_profiles/ny_freeway_to _boulvard_rochester.aspx.

204 *Eastman Kodak was founded:* "Eastman Kodak Company," *Democrat and Chronicle,* May 24, 1892, www.newspapers.com/clip/31921321/may-23-1892/.

204 *during its heyday:* Kaitlyn Tiffany, "The Rise and Fall of an American Tech Giant," *Atlantic,* June 16, 2021, www.theatlantic.com/magazine/archive/2021/07/kodak -rochester-new-york/619009/; Census Tracts: Rochester, New York, U.S. Department of Commerce (1970).

204 *spread his wealth:* "Kodak History: George Eastman," www.kodak.com/en/ company/page/george-eastman-history.

204 *Home Owners' Loan Corporation:* Robert K. Nelson et al., "Mapping Inequality," in Nelson and Ayers, *American Panorama,* dsl.richmond.edu/panorama/redlining/ #loc=12/43.188/-77.697&city=rochester-ny.

204 *Inner Loop highway:* "City Studies Loop Theory on Traffic," *Democrat and Chronicle,* Sept. 16, 1948.

205 *Opposition to such a project:* "City Cold to Plea Inner Loop Spare 3d Ward Shrines," *Democrat and Chronicle,* July 12, 1952.

205 *updates on the highway's progress:* "Route of Inner Loop," *Rochester Times-Union,* July 25, 1952; Truman Searle, "Are You Up-to-Date on Our Loops?," *Democrat and Chronicle,* July 12, 1953.

205 *the first downtown loops:* "Downtown Loop First in State," *Rochester Times-Union,* Sept. 22, 1965.

205 *a story of demolition:* "Last Home Removed in Path of Loop Arc," *Democrat and Chronicle,* Aug. 4, 1956.

205 *Eva Woods was living:* "A Moving Plea Clears Way for Inner Loop Section," *Rochester Times-Union,* May 11, 1962.

206 *More than a thousand:* "Inner Loop North Transformation Planning Study," City of Rochester, Aug. 2022, 12.

206 *ribbon-cutting ceremony:* "Loop Booming Downtown Business: Rocky," *Rochester Times-Union,* Oct. 20, 1965.

206 *population would peak:* Sean Lahman, "Rochester Population Falls Out of Top 100," *Democrat and Chronicle,* May 21, 2015, www.democratandchronicle.com/story/ news/2015/05/21/rochester-population-falls-top/27710675.

206 *"chokes downtown Rochester":* Frank Zoretich, "Inner Loop—Costly Roadway of Ugliness," *Democrat and Chronicle,* Aug. 24, 1972.

206 *a comprehensive plan:* "Inner Loop North: About," City of Rochester, www
.innerloopnorth.com/about; "Rochester 2010: The Renaissance Plan," City of
Rochester Community Development, Bureau of Planning, Jan. 1999, 82.

206 *seven thousand cars a day:* "Inner Loop East Reconstruction Project," application for
FY 2013 TIGER Discretionary Grant Program, June 3, 2013, 4.

207 *eighteen feet belowground:* "Loop's Big Headaches Are Under the Ground," *Democrat
and Chronicle,* Nov. 22, 1956.

208 *six acres of developable land:* "Inner Loop East Project," City of Rochester, www
.cityofrochester.gov/InnerLoopEast/.

208 *downtown was growing:* "Downtown Housing Market Report," Rochester Downtown Development Corporation, Dec. 2018, 5.

209 *More than $300 million:* James Brown, "Can Biden's Infrastructure Plan Bury the
Inner Loop for Good? City Hall Hopes So," *CITY,* May 27, 2021, www.roccitynews
.com/news-opinion/can-bidens-infrastructure-plan-bury-the-inner-loop-for-good
-city-hall-hopes-so-13243996.

209 *more than five hundred:* "Inner Loop North Transformation Planning Study," City of
Rochester, Aug. 2022, 57.

210 *the city began to study:* Ibid., 20.

210 *Chuck Schumer visited Rochester:* "Senators Schumer, Gillibrand Discuss Inner Loop
North Project in Rochester," WROC, June 29, 2021, www.youtube.com/watch?v=
Kfknr3rlDtI.

213 *six concepts for removing:* "Inner Loop North Transformation Planning Study," 99.

213 *announced its preferred concept:* "Rochester 2034: Moving Forward Progress Report,"
City of Rochester Office of Planning, Nov. 2021, 26.

213 *included $100 million:* Randy Gorbman, "Hochul Announces $100 Million Commitment for Inner Loop Fill-In Work," WXXI News, March 7, 2022, www.wxxinews
.org/local-news/2022-03-07/hochul-announces-100-million-commitment-for-inner
-loop-fill-in-work.

216 *presented a plan:* "Lewis Street Presentations—Hinge Neighbors," April 17, 2021,
hingeneighbors.com/neighborhood-presentations.

Chapter 17: Land

217 *Sanborn Map Company:* Walter W. Ristow, "Fire Insurance Maps: In the Library of
Congress," Library of Congress, www.loc.gov/collections/sanborn-maps/articles
-and-essays/introduction-to-the-collection/.

217 *1940 Sanborn map of Austin:* Sanborn Fire Insurance Map from Austin, Travis
County, Texas, Sanborn Map Company, vol. 3, 1940, Texas State Library and Archives Commission.

218 *bought the property in 1999:* Travis County Appraisal District Real Property Account
Information, Property No. 188235.

218 *started drawing up plans:* "Demolition and Relocation Permits," Austin Historic
Landmark Commission, April 22, 2019.

218 *TxDOT planned to take:* "Draft Environmental Impact Statement: I-35 Capital Express Central Project from US 290 East to US 290 West/SH 71," 127.

219 *His realization was prescient:* Carl Hedman et al., "Austin and the State of Low- and
Middle-Income Housing," Urban Institute, Oct. 2017, 113–14; "East Austin, TX Median Sales Price," Redfin calculation of home data from MLS, www.redfin.com/
neighborhood/351579/TX/Austin/East-Austin/housing-market.

219 *The first community land trust:* John Emmeus Davis, ed., *The Community Land Trust
Reader* (Cambridge, Mass.: Lincoln Institute of Land Policy, 2010), 3.

219 *movement of sharecroppers:* Ibid., 194, cltroots.org/interviews/charles-sherrod
-interview.

220 *co-founded New Communities:* Davis, *Community Land Trust Reader,* 16–17.

220 *applied for an emergency loan:* Debbie Weingarten, "How a Black Farming Community Found Justice," Civil Eats, Aug. 16, 2019, civileats.com/2019/08/16/how-a
-black-farming-community-found-justice/.

220 *increasingly urban phenomenon:* James Meehan, "Reinventing Real Estate: The Community Land Trust as a Social Invention in Affordable Housing," *Journal of Applied
Social Sciences* 8, no. 2 (2014): 123.

220 *state's property code:* Local Government Code, Title 12, Subtitle A, Chapter 373B:
Community Land Trusts, statutes.capitol.texas.gov/Docs/LG/htm/LG.373B.htm.

220 *first community land trust home:* Eliza Platts-Mills, "A Guide for Developing Community Land Trust Affordable Homeownership Programs in Texas," Entrepreneurship
and Community Development Clinic, University of Texas School of Law (2018), 7.

222 *Mexican-owned pharmacies:* "Zoning Change Review Sheet," City of Austin (2011),
Case No. C14H-2011-0002.

223 *assembling eight acres:* Kevin Fullerton, "Naked City: Back on Tract," *Austin Chronicle,* Sept. 1, 2000, www.austinchronicle.com/news/2000-09-01/78468/.

223 *produced a documentary:* Andrew Cadelago, Lauren Hardy, and Isaac Simon, "La
Casa Lopez," East Austin Stories, University of Texas at Austin Moody College of
Communication, rtf.utexas.edu/east-austin-stories.

224 *zoned the Lopez house:* "Zoning Change Review Sheet."

224 *GNDC submitted plans:* "Guadalupe Neighborhood Development Corporation
Rental Housing Development Assistance Application," Nov. 6, 2020.

224 *La Vista de Lopez:* James Rambin, "At La Vista de Lopez, an East Austin Holdout
Imagines a Tower of Its Own," Towers.net, Aug. 7, 2019, austin.towers.net/at-la
-vista-de-lopez-an-east-austin-holdout-imagines-a-tower-of-its-own/.

224 *"mending the divide":* "Our Future 35—Austin's Cap and Stitch Program," www
.ourfuture35.com.

224 *the cost of capping:* Kirk Watson, "Item 44," City of Austin Council Message Board,
Feb. 20, 2023, austincouncilforum.org/viewtopic.php?t=1750.

224 *roughly fifteen acres:* James Rambin, "What Goes on Top of a Highway?" Towers
.net, Jan. 20, 2023, austin.towers.net/what-goes-on-top-of-a-highway/.

Chapter 18: Leverage

229 *black-and-white images:* "Advocates Briefing to Harris County Attorney," April 30,
2021, 4–5.

232 *Our Afrikan Family:* "About Us," ourafrikanfamily.org.

237 *has been demolished:* John Wayne Ferguson, "Stop I-45 Group Cries Foul over Re-
started Demolitions of Lofts at the Ballpark," *Houston Chronicle,* Sept. 2, 2022, www
.houstonchronicle.com/news/houston-texas/housing/article/Stop-I-45-group
-cries-foul-over-restarted-17416778.php.

237 *gave a tour:* Email from Allen Douglas (executive director Houston Downtown Redevelopment Authority) to Bruce Rychlik (Texas Department of Transportation),
July 8, 2022.

238 *TxDOT confirmed publicly:* Shelley Childers, "Neighbors, 'Stop TxDOT I-45' Fighting for Remaining Parts of EaDo Housing Complex," KTRK, Oct. 18, 2022, abc13
.com/txdot-i-45-expansion-eado-apartments-demolitions/12343109/.

238 *two dozen distinct watersheds:* "About the Flood Control District," Harris County
Flood Control District, www.hcfcd.org/About/About-the-Flood-Control-District.

238 *"any program or project":* "Section 4(f)," *Public Roads,* Federal Highway Administration, June 27, 2017, www.fhwa.dot.gov/infrastructure/50section4f.cfm.

239 *support for a petition:* Fred Linder, "Create White Oak Bayou Park," Action Network, actionnetwork.org/petitions/create-white-oak-bayou-park.

239 *wouldn't necessarily halt:* "The Challenges of Naming White Oak Bayou Greenway as a City Park," letter from Karla Cisneros, Houston City Council Member, Jan. 10, 2023.

239 *"important news conference":* Jay R. Jordan (@JayRJordan), Twitter, Dec. 18, 2022, 7:09 P.M., twitter.com/jayrjordan/status/1604694350646153222?s=20.

240 *"noble locomotive":* Betty T. Chapman, "Plow and Locomotive on City Seal Illustrate Origins of Houston," *Houston Business Journal,* July 14, 1995, www.houstontx.gov/abouthouston/cityseal.pdf.

240 *"I've always said":* "City of Houston, Harris County, and TxDOT Announce New Memorandum of Understanding for the North Houston Highway Improvement Project," City of Houston Mayor's Office, Dec. 19, 2022, www.houstontx.gov/mayor/press/2022/nhhip-mou.html.

240 *the city's MOU:* Memorandum of Understanding Between the Texas Department of Transportation and the City of Houston Concerning the North Houston Highway Improvement Project, signed Dec. 19, 2022, ftp.txdot.gov/pub/txdot/hou/nhhip/mou-txdot-houston-nhhip.pdf.

240 *Harris County's agreement:* Memorandum of Understanding Between the Texas Department of Transportation and Harris County Concerning the North Houston Highway Improvement Project, signed Dec. 22, 2022, www.txdot.gov/content/dam/project-sites/nhhip/docs/harris-county-mou-12-19-2022.pdf.

Chapter 19: Forward

245 *local painter named Ellwood Blackman:* Mark Stuertz, "The Living End," *Dallas Observer,* Oct. 14, 2004, www.dallasobserver.com/news/the-living-end-6420079.

246 *the parade begins:* Sarah Bahari, "Community Gathers to Honor Civil Rights Leader at Dallas' First MLK Parade in 3 Years," *Dallas Morning News,* Jan. 16, 2023, www.dallasnews.com/news/2023/01/16/community-gathers-to-honor-civil-rights-leader-at-dallas-first-mlk-parade-in-3-years/.

249 *six pages into the consent agenda:* Dallas City Council Agenda (Revised), Feb. 22, 2023, 7.

249 *the resolution had been pulled:* Matt Goodman, "The I-345 City Council Vote Will Not Be Happening Next Week," *D Magazine,* Feb. 14, 2023, www.dmagazine.com/frontburner/2023/02/the-i-345-city-council-vote-will-not-be-happening-next-week/.

249 *wrote a memo:* Matt Goodman, "Five Dallas Council Members Call for Independent Study of I-345 Removal," *D Magazine,* April 10, 2023, www.dmagazine.com/frontburner/2023/04/five-dallas-council-members-call-for-independent-study-of-i-345-removal/.

249 *TxDOT had recently spent $30 million:* "I-345 Feasibility Study," Texas Department of Transportation, Aug. 2022, 4.

250 *council convenes:* Dallas City Council, Public Meeting, May 24, 2023, dallastx.new.swagit.com/videos/232304.

250 *"Dallas is a very young city":* Matt Goodman, "I-345 Is Dead, Long Live I-345," *D Magazine,* May 24, 2023, www.dmagazine.com/frontburner/2023/05/i-345-dallas-city-council-vote-hybrid-plan-txdot/.

251 *"After a dozen years":* Ibid.

Chapter 20: Fight

253 *opened the Whip In:* James McWilliams, "How a Texas Convenience Store Became a Locavore Haven," *Atlantic,* Oct. 13, 2010, www.theatlantic.com/health/archive/2010/10/how-a-texas-convenience-store-became-a-locavore-haven/64461/.

253 *its initial environmental review:* "Draft Environmental Impact Statement: I-35 Capital Express Central Project from US 290 East to US 290 West/SH 71."

254 *preferred design:* Ibid., S-6.

254 *Along eight miles:* Ibid., S-7.

254 *a local TV reporter:* Tahera Rahman, "'I Wouldn't Know What to Do': Business Finds Out It's in Potential I-35 Displacement Path," KXAN, Jan. 4, 2023, www.kxan.com/news/local/austin/i-wouldnt-know-what-to-do-business-finds-out-its-in-potential-i-35-displacement-path.

254 *The environmental impact statement:* "Draft Environmental Impact Statement: I-35 Capital Express Central Project from US 290 East to US 290 West/SH 71."

254 *greenhouse gas analysis:* "Draft Environmental Impact Statement: I-35 Capital Express Central Project from US 290 East to US 290 West/SH 71, Appendix V: Greenhouse Gas Analysis and Climate Change Assessment," Texas Department of Transportation, Nov. 2022, ES-14.

254 *In 2017, when TxDOT looked at:* "Draft Environmental Impact Statement: North Houston Highway Improvement Project," Texas Department of Transportation, April 2017, 4-1 to 4-3.

255 *that benefit flattens:* "Statewide On-Road Greenhouse Gas Emissions Analysis and Climate Change Assessment," Texas Department of Transportation, 11.

255 *In the I-35 study:* "Draft Environmental Impact Statement: I-35 Capital Express Central Project," 423.

255 *induced travel generated:* Ibid., 425, table 3.24-1.

255 *an independent analysis:* Rocky Mountain Institute SHIFT Calculator: State Highway Induced Frequency of Travel, shift.rmi.org.

255 *average daily traffic:* "Draft Environmental Impact Statement: I-35 Capital Express Central Project," 8–10.

256 *Since 2000, daily traffic counts:* Jack Craver, "Why Are We Expanding I-35 Again?," *Austin Politics Newsletter,* mailchi.mp/306368ecaa2c/apn-14-why-are-we-expanding-i-35-again?e=a1c3c89bd6.

256 *Los Angeles County:* Rachel Urgana, "710 Freeway Expansion Dropped After Decades of Planning, Marking a Milestone for L.A.," *Los Angeles Times,* May 26, 2022, www.latimes.com/california/story/2022-05-26/710-freeway-expansion-los-angeles-plan-killed.

256 *Colorado Department of Transportation:* Michael Booth, "A Lane Expansion to Unclog I-25 Through Downtown Denver Is Not on the Table—for Now," *Colorado Sun,* May 16, 2022, coloradosun.com/2022/05/16/i-25-no-expansion-central-denver/.

256 *"People don't think of ODOT":* Laura Bliss, "The Road Warriors," Bloomberg, Jan. 22, 2022, www.bloomberg.com/news/features/2022-01-22/in-portland-youth-activists-are-driving-a-highway-revolt?sref=0IejgNtz.

256 *rescinded its approval:* Phillip A. Ditzler (FHWA division administrator) to Kris Stricker (executive director, Oregon Department of Transportation), Jan. 18, 2022.

257 *A group of students and faculty:* Madeline Duncan, "'Completely Absurd': Student Organization to Challenge I-35 Expansion to 20 Lanes," *Daily Texan,* Sept. 20, 2022, thedailytexan.com/2022/09/20/completely-absurd-student-organization-to-challenge-i-35-expansion-to-20-lanes/.

257 *Israel told Rethink35:* "Candidates' Positions on I-35: Celia Israel," Rethink35.com, accessed via Wayback Machine, web.archive.org/web/20221018224505/https://rethink35.com/candidates-positions-on-i-35/celia-israel.

257 *"Whether you like cars":* "Candidates' Positions on I-35: Kirk Watson," Rethink35.com, accessed via Wayback Machine, web.archive.org/web/20221019102433/https:/rethink35.com/candidates-positions-on-i-35/kirk-watson.

257 *nine hundred votes:* Joshua Fletcher, "Austin Voters Elect Kirk Watson, Who Served

as Mayor Two Decades Ago, to Lead the City Again," *Texas Tribune*, Dec. 13, 2022, www.texastribune.org/2022/12/13/austin-mayor-runoff-kirk-watson-celia-israel/.

259 *the agency holds a public hearing:* "I-35 Capital Express Central Public Hearing," Texas Department of Transportation, Feb. 9, 2023, my35capex.com/events/i-35 -capital-express-central-public-hearing/.

259 *TxDOT announces:* "I-35 Capital Express Central Project Public Hearing Exhibits," Texas Department of Transportation, Feb. 9, 2023, 24, my35capex.com/draft-eis/.

259 *a list of demands:* Nathan Bernier, "Austin City Council Demands Changes to I-35 Plan," KUT, Feb. 23, 2023, www.kut.org/transportation/2023-02-23/txdot-i-35 -capital-express-project.

259 *Although many council members:* Austin City Council, Resolution No. 20230223-044, adopted Feb. 23, 2023.

259 *opens the meeting:* Austin City Council Public Meeting, Item 44, Feb. 23, 2023, austintx.new.swagit.com/videos/208896.

260 *vote against the resolution:* Bernier, "Austin City Council Demands Changes to I-35 Plan."

260 *joined the Climate Mayors coalition:* "Climate Mayors: Austin, Kirk Watson," climatemayors.org/city/austin/.

260 *note to his colleagues:* Kirk Watson, "Item 44," City of Austin Council Message Board, Feb. 20, 2023, austincouncilforum.org/viewtopic.php?t=1750.

261 *announces the forty-five projects:* "Biden-Harris Administration Announces First-Ever Awards from Program to Reconnect Communities," U.S. Department of Transportation, Feb. 28, 2023.

261 *$1.1 million planning grant:* "Our Future 35: Connecting Austin Equitably—Mobility Study, Reconnect Communities Pilot Program FY 2022 Award Fact Sheets," U.S. Department of Transportation, Feb. 28, 2023.

261 *holds a groundbreaking ceremony:* Nathan Bernier, "Construction Starts in North Austin on Five-Year Project to Expand I-35," KUT, March 29, 2023, www.kut.org/2023 -03-29/construction-starts-in-north-austin-on-five-year-project-to-expand-i-35.

261 *Youth Day of Action:* "Judah Rice Speaking at the UT Austin I-35 Youth Day of Action," Rethink35, Feb. 15, 2023, www.youtube.com/watch?v=4nkAkbtX8PU.

Chapter 21: Resolution

263 *commission will begin discussion:* Texas Transportation Commission public meeting, Feb. 23, 2023, txdot.new.swagit.com/videos/208914.

263 *"This 10-year plan":* "Governor Abbott, TxDOT Announce Record $100 Billion 10-Year Transportation Plan," Office of the Texas Governor, Feb. 23, 2023, gov.texas .gov/news/post/governor-abbott-txdot-announce-record-100-billion-10-year -transportation-plan.

263 *flush with cash:* Reese Oxner, "Biden's Infrastructure Plan Will Set Aside About $35 Billion for Texas Projects," *Texas Tribune*, Nov. 9, 2021, www.texastribune.org/ 2021/11/09/biden-infrastructure-bill-texas/.

264 *voted to delay funding:* Houston-Galveston Area Council Transportation Policy Council public meeting, Item 8, Jan. 27, 2023, hgac.swagit.com/play/01272023 -609.

264 *when Houston joined:* "United Bid Committee Starts Outreach for FIFA 2026 WC Host Cities," U.S. Soccer, Aug. 15, 2017, www.ussoccer.com/stories/2017/08/ united-bid-committee-commences-outreach-for-potential-host-cities-in-bid-for-2026 -fifa-world-cup.

264 *massive improvement campaign:* Dug Begley, "World Cup Could Cap Off Plans for Downtown Park atop I-45 Project," *Houston Chronicle*, June 14, 2018, www

.houstonchronicle.com/news/houston-texas/houston/article/World-Cup-could
-cap-off-plans-for-downtown-park-12992698.php.

265 *wrote to Paul:* Email from Kristopher Larson (president and CEO, Central Houston Inc.) to Eliza Paul and Varuna Singh (TxDOT Houston deputy district engineer), June 23, 2022.

265 *the cost of highway construction:* Jeff Davis, "Highway Construction Costs Have Risen 50% in Two Years," Eno Center for Transportation, April 18, 2023, www .enotrans.org/article/highway-construction-costs-have-risen-50-in-two-years/.

266 *Congress confirmed:* "U.S. Department of Transportation Announces Confirmation of Shailen Bhatt as 21st Administrator of the Federal Highway Administration," Federal Highway Administration, Dec. 8, 2022, highways.dot.gov/newsroom/us-department -transportation-announces-confirmation-shailen-bhatt-21st-administrator.

267 *follows the same itinerary:* "Congresswoman Sheila Jackson Lee Hosts Federal High-way Administrator Shailen Bhatt in Houston Today on I-45 Project and Federal Funding," Congresswoman Sheila Jackson Lee, Facebook live, www.facebook .com/CongresswomanSheilaJacksonLee/videos/1452889212184884, www .facebook.com/CongresswomanSheilaJacksonLee/videos/991781581807003.

267 *rebuilt their church:* Keri Blakinger, "Houston Church Opens Again After Hurricane Ike Destruction," Chron.com, Oct. 23, 2016, www.chron.com/news/houston -texas/article/Houston-church-opens-again-after-Hurricane-Ike-10154565.php.

267 *the group reconvenes:* Taisha Walker, "Federal Highway Administrator, TxDOT Fi-nalize $9 Billion I-45 Expansion Project," KPRC 2, March 16, 2023.

268 *voluntary resolution agreement:* "FHWA and TxDOT Sign Agreement to Allow I-45 North Houston Highway Improvement Project to Move Forward," Texas Depart-ment of Transportation, March 7, 2023, www.txdot.gov/about/newsroom/local/ houston/fhwa-txdot-sign-agreement-i45-nhhip.html.

269 *monitoring the project:* Ally Smither, "FHWA Meeting with Stakeholders & Commu-nity Members 5–6PM," March 6, 2023.

269 *TxDOT had committed:* "Voluntary Resolution Agreement," signed March 6, 2023.

269 *Modesti Cooper listened:* Smither, "FHWA Meeting with Stakeholders."

269 *required adding capacity:* "About This Project," NHHIP, Texas Department of Trans-portation, www.txdot.gov/nhhip/about.html.html.

270 *FHWA sent a letter:* Achille Alonzi (FHWA division administrator) to Williams, March 6, 2023.

Chapter 22: Proximity

271 *enabling seamless car travel:* "Understanding the 2021 Infrastructure Law," Transpor-tation for America, t4america.org/iija/#start.

271 *Biden wanted two-thirds:* Coral Davenport, "E.P.A. Lays Out Rules to Turbocharge Sales of Electric Cars and Trucks," *New York Times,* April 12, 2023, www.nytimes .com/2023/04/12/climate/biden-electric-cars-epa.html.

271 *committed $7.5 billion:* "A Guidebook to the Bipartisan Infrastructure Law," White House, www.whitehouse.gov/build/guidebook.

271 *a Ford F-150 Lightning:* Maegan Vazquez and Jason Hoffman, "Biden Takes New Electric F-150 for a Test Drive: 'This Sucker's Quick,'" CNN, May 18, 2021, www .cnn.com/2021/05/18/politics/biden-ford-test-drive/index.html.

272 *electric GMC Hummer:* David Zipper, "Electric Vehicles Are Bringing Out the Worst in Us," *Atlantic,* Jan. 4, 2023, www.theatlantic.com/ideas/archive/2023/01/electric -vehicles-suv-battery-climate-safety/672576/.

272 *"the fleet turnover time":* Costa Samaras (@CostaSamaras), Twitter, Sept. 15, 2020, 11:04 P.M., threadreaderapp.com/thread/1306066319129878529.html; Abdullah F. Alarfaj, W. Michael Griffin, and Constantine Samaras, "Decarbonizing US Passen-

ger Vehicle Transport Under Electrification and Automation Uncertainty Has a Travel Budget," *Environmental Research Letters* 15 (2020).

272 *2.75 trillion miles:* "Annual Vehicle Miles Traveled in the United States," Alternative Fuels Data Center, U.S. Department of Energy, afdc.energy.gov/data/10315.

272 *vehicle emissions jumped:* "Driving Down Emissions," Transportation for America, Oct. 2020, 11.

272 *reduce carbon emissions:* Ben Holland et al., "Urban Land Use Reform: The Missing Key to Climate Action," Rocky Mountain Institute (2023), rmi.org/insight/urban -land-use-reform/.

273 *more dense residential development:* "Land Development Code Revision: Report Card," City of Austin, Feb. 21, 2020.

273 *increase greenhouse gas emissions:* "Issue Brief: Estimating the Greenhouse Gas Im- pact of Federal Infrastructure Investments in the IIJA," Georgetown Climate Cen- ter, Dec. 16, 2021, www.georgetownclimate.org/articles/federal-infrastructure -investment-analysis.html.

273 *"minimizing further highway expansion":* James Bradbury et al., "Issue Brief: States Are in the Driver's Seat on Transportation Carbon Pollution," Georgetown Climate Center, March 24, 2023, www.georgetownclimate.org/blog/states-in-the-driver-eys -seat.html.

273 *survey to freeway fighters:* "Survey of Community Connectors," Transportation for America, Jan. 2023.

274 *less than 4 percent:* Helene Desanlis et al., "Funding Trends 2021: Climate Change Mitigation Philanthropy," Climateworks Global Intelligence, Oct. 2021.

274 *"Already the automobile":* Norman Bel Geddes, *Magic Motorways* (New York: Random House, 1940), 10.

275 *"One of the best ways":* Ibid., 6.

275 *A dozen states:* "Freeway Fighting Projects Across the United States: Expansion Pre- vention Campaigns," Freeway Fighters Network, freeway-fighters.org.

275 *"If annual appropriations":* Shannon, "Untrustworthy Highway Fund."

276 *"a transportation financing system":* Congressional Record, U.S. Senate Committee on Public Works, Jan. 18, 1973, 88–92.

276 *House Bill 1261:* "Climate Action Plan to Reduce Pollution," Colorado General As- sembly, 2019 regular session, leg.colorado.gov/bills/hb19-1261.

276 *Colorado's transportation commission:* Greenhouse Gas (GHG) Program: GHG Transportation Planning Standard, Colorado Department of Transportation, ap- proved Dec. 16, 2021, www.codot.gov/programs/environmental/greenhousegas.

276 *adding four lanes:* Booth, "Lane Expansion to Unclog I-25 Through Downtown Den- ver Is Not on the Table—for Now."

276 *remove an elevated stretch of Interstate 81:* Michelle Breidenbach, "Here's What the First Part of Syracuse's I-81 Construction Will Look Like (Maps)," Syracuse.com, June 2, 2022, www.syracuse.com/news/2022/06/heres-what-the-first-part-of -syracuses-i-81-construction-will-look-like-maps.html.

276 *"You just have to step back":* "Gov. Kathy Hochul on I-81 in Syracuse: 'What Were They Thinking?,'" Syracuse.com, Jan. 25, 2022, www.youtube.com/watch?v= JjCCBP0u4z0.

276 *to remove I-375:* Ian Duncan, "Detroit Wins Grant to Remove Interstate That Wrecked a Black Community," *Washington Post,* Sept. 15, 2022, www .washingtonpost.com/transportation/2022/09/15/detroit-highway-removal -paradise-valley/.

277 *"We don't talk about":* Melissa Nann Burke and Riley Beggin, "Michigan Gets $105M Grant from Feds to Turn I-375 in Detroit into Boulevard," *Detroit News,* Sept. 15, 2022, www.detroitnews.com/story/news/politics/2022/09/15/michigan-gets-105 -m-federal-grant-turn-375-detroit-into-boulevard/10380105002/.

277 *proposals to remove:* Lauren Mayer, "Freeways Without Futures 2023," Congress for the New Urbanism.

277 *marking fifteen years:* "Freeways Without Futures," Congress for the New Urbanism, www.cnu.org/our-projects/highways-boulevards/freeways-without-futures.

277 *For the first time:* Mayer, "Freeways Without Futures 2023," 4.

277 *what about building a park:* Megan Kimble, "Can Anacostia Build a Bridge Without Displacing Its People?," *New York Times,* Aug. 10, 2022, www.nytimes.com/interactive/2022/08/09/headway/anacostia-bridge.html.

280 *In August, TxDOT published:* "Final Environmental Impact Statement and Record of Decision: I-35 Capital Express Central Project from US 290 East to US 290 West/SH 71," Texas Department of Transportation, August 2023.

280 *TxDOT claimed adding:* "Final Environmental Impact Statement and Record of Decision: I-35 Capital Express Central Project from US 290 East to US 290 West/SH 71, Appendix V: Greenhouse Gas and Climate Change," 31.

281 *Widening the highway would generate:* "Mapped: The World's Coal Power Plants," Carbon Brief, March 26, 2020, www.carbonbrief.org/mapped-worlds-coal-power-plants.

282 *nearly two decades to complete:* Dug Begley, "Nearly Two Decades: The Long, Long Journey Ahead for $9.7B Rebuild of I-45," *Houston Chronicle,* May 7, 2023, www.houstonchronicle.com/news/houston-texas/transportation/article/harris-county-i-45-rebuilding-project-houston-18073985.php?sid=5fb407d43bc47e5bce5e04b8&ss=A&st_rid=null&utm_source=newsletter&utm_medium=email&utm_term=news_a&utm_campaign=HC_AfternoonReport.

283 *Fair for Houston:* Yilun Cheng, "Advocates Say Houston Is Underrepresented on Regional Council. They Have a Petition to Change That," *Houston Chronicle,* March 3, 2023, www.houstonchronicle.com/politics/houston/article/petition-to-remove-houston-from-regional-council-17817879.php.

283 *Houston and Harris County made up:* Fair For Houston, www.fairforhouston.com.

284 *when the council allocated:* Dylan McGuinness, "Houston Set to Get Just 2% of $488M in Federal Flood Funds from Regional Council," *Houston Chronicle,* Feb. 15, 2022, www.houstonchronicle.com/news/houston-texas/houston/article/Houston-once-again-feels-stiffed-on-federal-16921712.php.

284 Ramps to Nowhere: Minda Martin, *Ramps to Nowhere* (2018), vimeo.com/361862625.

Epilogue

287 *711 acres of land:* Mike Clarke-Madison, "Mulling over Mueller," *Austin Chronicle,* Oct. 31, 1997, www.austinchronicle.com/news/1997-10-31/518698/.

287 *25 percent of homes:* "Master Development Agreement Between the City of Austin and Catellus Austin, LLC," Dec. 2, 2004, 60.

289 *the city published renderings:* Austin Sanders, "Taking a Look at the Neighborhoods That Will Be Changed the Most by Project Connect," *Austin Chronicle,* June 3, 2022.

INDEX

ABOUT THE AUTHOR

MEGAN KIMBLE is an investigative journalist and the author of *Unprocessed*. A former executive editor at *The Texas Observer*, Kimble writes about housing, transportation, and urban development for *The New York Times*, *Texas Monthly*, *The Guardian*, and Bloomberg CityLab. She lives in Austin, Texas.

megankimble.com

ABOUT THE TYPE

This book was set in Dante, a typeface designed by Giovanni Mardersteig (1892–1977). Conceived as a private type for the Officina Bodoni in Verona, Italy, Dante was originally cut only for hand composition by Charles Malin, the famous Parisian punch cutter, between 1946 and 1952. Its first use was in an edition of Boccaccio's *Trattatello in laude di Dante* that appeared in 1954. The Monotype Corporation's version of Dante followed in 1957. Though modeled on the Aldine type used for Pietro Cardinal Bembo's treatise *De Aetna* in 1495, Dante is a thoroughly modern interpretation of that venerable face.